AF541759

ENGINEERING MECHANICS

By

Shashi Kant Yadav

DISCOVERY PUBLISHING HOUSE
NEW DELHI-110002

First Published - 2006
Reprinted - 2016

ISBN: 978-81-8356-108-2

Engineering Mechanics

Published by:
DISCOVERY PUBLISHING HOUSE PVT. LTD.
4383/4B, Ansari Road, Darya Ganj
New Delhi-110 002 (India)
Phone: +91-11-23279245, 43596064-65
Fax: +91-11-23253475
E-mail: discoverypublishinghouse@gmail.com
sales@discoverypublishinggroup.com
web: www.discoverypublishinggroup.com

Printed at:
Infinity Imaging Systems
Delhi

Preface

The present title "Engineering Mechanics" has been written for the undergraduate and those preparing for the higher national certificate and professional institution examinations, as well as for these following a degree, or diploma course. The main aim has been to give a clear understanding of the principles underlying engineering design, and a special effort has been made to indicate the shortest analysis of a wide variety of problems. Each chapter is complete in itself and is built up logically to cover all aspects of the particular theory. The book is written in a simple and easy to follow language, so that even an average student can grasp the subject by self study.

In the preparation of this book large number of books and research papers have been consulted. So no authenticity is claimed.

The author wishes to express his deepest appreciation to the many people who have contributed in one way or the other to the preparation of this title.

The author expresses his gratitude to Mr. Wasan and staff of M/s Discovery Publishing House for their whole hearted co-operation in the publication of this book.

The author tried hard to be accurate and upto date in statement and realises the impossibility of completely avoiding errors therefore, the author will greatly appreciate having his attention called to any questionable statement.

Author

Preface

The present title "Engineering Mechanics" has been written for the undergraduate and those preparing for the higher national certificate and professional institution examinations, as well as for those following a degree, or diploma course. The main aim has been to give a clear understanding of the principles underlying engineering design, and a special effort has been made to indicate the shortest analysis of a wide variety of problems. Each chapter is complete in itself and is built up logically to cover all aspects of the particular theory. The book is written in a simple and easy to follow language, so that even an average student can grasp the subject by self study.

In the preparation of this book large number of books and research papers have been consulted, so no authenticity is claimed.

The author wishes to express his deepest appreciation to the many people who have contributed in one way or the other to the preparation of this title.

The author expresses his gratitude to Mr. Wasan and staff of M/s Discovery Publishing House for their whole hearted co-operation in the publication of this book.

The author tried hard to be accurate and upto date in statement and realises the impossibility of completely avoiding errors therefore, the author will greatly appreciate having his attention called to any questionable statement.

Author

Contents

1

Fundamentals of Engineering Mechanics

INTRODUCTION

That branch of science which deals with the behaviour of a body when the body is at rest or in motion is called engineering mechanics. Statics and Dynamics are the two major of engineering mechanics. Statics is the branch of sciences that deals with the study of a body when the body is at rest. Whereas dynamics referes to that branch of science which deals with the study of a body when the body is in motions Dynamics is further divided into kinetics. When the forces which cause the motion are not considered in the study of a body in motion, then it is called kinematics. Similarly kinetics refers to the forces that are cosnidered for the body in motion. Figure 1.1 shows the classification of Engineering Mechanics.

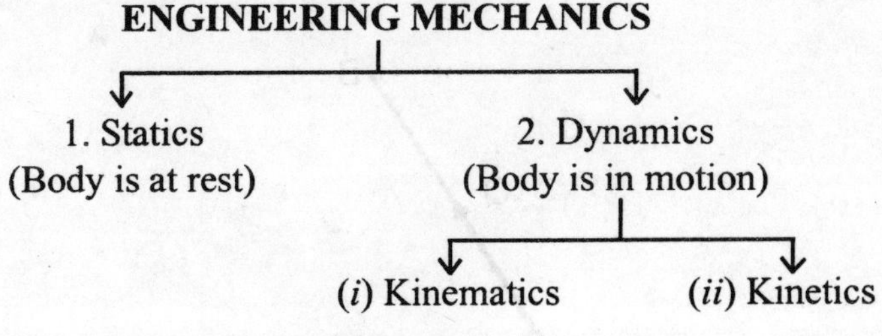

Figure 1.1

DEFINITIONS

Vector Quantity : Vector quantity is a quantity which is completely specitied by magnitude and direction. Velocity, acceleration, force and momentum are a few examples of vector quantity. A vector quantity is represented by means of a straight line with an arrow. The length of the straight line (i.e., *AB*) represents the magnitude and arrow represents the direction of the vector. The symbol $\overrightarrow{AB}$ also represents this vector, which means it is acting from *A* to *B*.

Figure 1.2 : *Vector Quantity.*

Scalar Quantity : Scalar quantity is a quantity which is completely specified by magnitude. Mass, length, time and temperature are a few examples of scalar quantity.

A Particle : A body of infinitely small volume, (or it particle is a body of negligible dimensions and the mass of the particle is considered to be concentrated at a point. Hence a particle is assumed to a point and the mass of the particle is concentrated at this point.

Law of Parallelogram of Forces: The law of parallelogram of forces is used to determine the resultant of two forces acting at a point in a plane. it states, two forces, acing at a point he represented in magnitude and direction by the two adjacent sides of a parallelogram, then their resultant is represented in magnitude and direction by the diagonal of the parallelogram passing through that point.

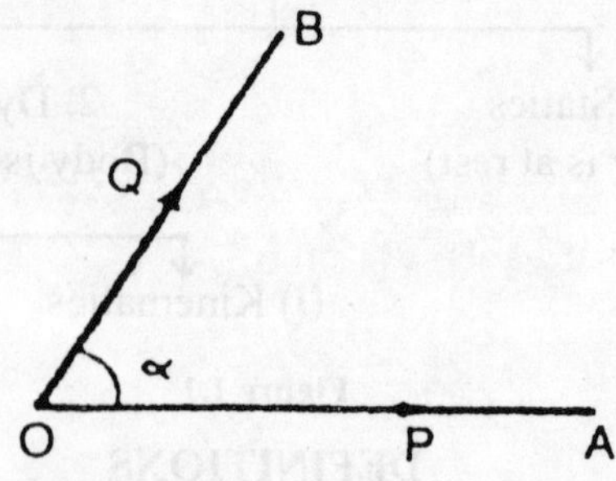

Figure : *1.3*

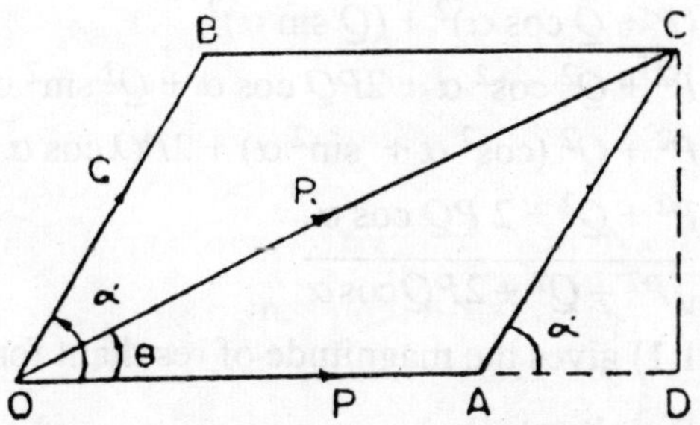

Figure : *1.4*

Let two forces *P* and *Q* act at a point *O* as shown in Figure 1.3. Thc forcc *P* is refpresented in magnitude and direction by *OA* whereas the force *Q* is presented in magnitude and direction by *OB*. Let the angle between the two forces by 'α'. The resultant of these two forces will be obtained in magnitude and idrection bythe diagonal (passing through *O*) of the parallelogram of which *OA* and *OB* are two adjacent sides. Hence draw the parallelogram with *OA* and *OB* as adjacent sides as shown in Figure 1.4. The resultant *R* is represented by *OC* iri magnitude and direction

Magnitude of Resultant (R)

From *C* draw *CD* perpendicular to *OA* produced.

Let α = Angle between two forces *P* and $Q = \angle AOB$

Now $\angle DAC = \angle AOB$ (Corresponding angles)

$= \alpha$

In parallelogram *OACB*, *AC* is parallel and equal to *OB*.

$\therefore \quad AC = Q.$

In triangle *ACD*,

$$AD = AC \cos \alpha = Q \cos \alpha$$

and $CD = AC \sin \alpha = Q \sin \alpha.$

In triangle *OCD*,

$$OC^2 = OD^2 + DC^2.$$

But $OC = R, OD = OA + AD = P + Q \cos \alpha$

and $DC = Q \sin \alpha.$

$$\therefore \quad R^2 = (P + Q\cos\alpha)^2 + (Q\sin\alpha)^2$$
$$= P^2 + Q^2\cos^2\alpha + 2PQ\cos\alpha + Q^2\sin^2\alpha$$
$$= P^2 + Q^2(\cos^2\alpha + \sin^2\alpha) + 2PQ\cos\alpha$$
$$= P^2 + Q^2 + 2PQ\cos\alpha$$

$$\therefore \quad R = \sqrt{P^2 + Q^2 + 2PQ\cos\alpha} \qquad (1.1)$$

Equation (1.1) gives the magnitude of resultant force R.

Direction of Resultant

Let θ = Angle made by resultant with OA.

Then from triangle OCD,

$$\therefore \quad \tan\theta = \frac{CD}{OD} = \frac{Q\sin\alpha}{P + Q\cos\alpha} \qquad (1.2)$$

$$\theta = \tan^{-1}\left(\frac{Q\sin\alpha}{P + Q\cos\alpha}\right)$$

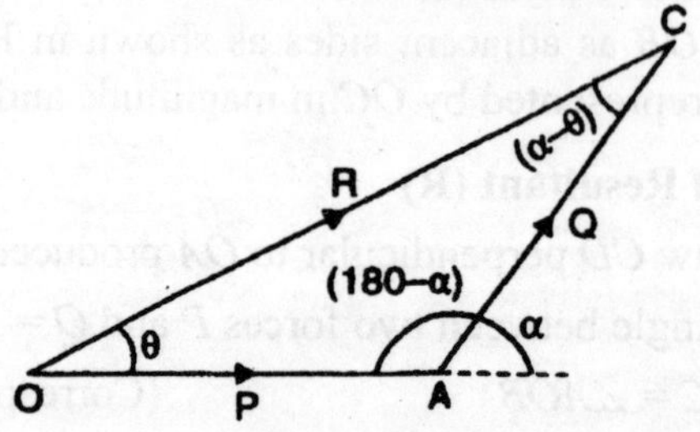

Figure 1.4 : *(a)*

By using sine rule the direction of resultant (R) can be obtained. [In triangle OAC, $OA = P$, $AC = Q$, $OC = R$, angle $OAC = (180 - \alpha)$, angle $ACO = 180 - [\theta + 180 - \alpha] = (\alpha - \theta)$]

$$\frac{\sin\theta}{AC} = \frac{\sin(180 - \alpha)}{OC} = \frac{\sin(\alpha - \theta)}{OA}$$

$$\frac{\sin\theta}{Q} = \frac{\sin(180 - \alpha)}{R} = \frac{\sin(\alpha - \theta)}{P}$$

Two cases are important.

I Case : When two forces P and Q are acting at right angles then

$$\alpha = 90°$$

we get the magnitude of resultant from Equation 1.1 as

$$R = \sqrt{P^2 + Q^2 + 2PQ\cos\alpha}$$

$$= \sqrt{P^2 + Q^2 + 2PQ\cos 90°}$$

$$= \sqrt{P^2 + Q^2} \qquad (\because \cos 90° = 0) \qquad (1.2\text{ A})$$

the direction of resultant is obtained from equation 1.2 as

$$\theta = \tan^{-1}\left(\frac{Q\sin\alpha}{P + Q\cos\alpha}\right)$$

$$= \tan^{-1}\left(\frac{Q\sin 90°}{P + Q\cos 90°}\right) = \tan^{-1}\frac{Q}{P} \quad (\because \sin 90° = 1 \text{ and } \cos 90° = 0)$$

II Case : When two equal forces P and Q are acting the angle between them is α then the magnitude and direction of result and is given as

$$R = \sqrt{P^2 + Q^2 + 2PQ\cos\alpha}$$

$$= \sqrt{P^2 + P^2 + 2P \times P \times \cos\alpha} \quad (\because P = Q)$$

$$= \sqrt{2P^2 \times 2\cos^2\frac{\alpha}{2}} \qquad \left(\because 1 + \cos\alpha = 2\cos^2\frac{\alpha}{2}\right)$$

$$= \sqrt{4P^2\cos^2\frac{\alpha}{2}} = 2P\cos\frac{\alpha}{2} \qquad (1.3)$$

and $\theta = \tan^{-1}\left(\dfrac{Q\sin\alpha}{P + Q\cos\alpha}\right) = \tan^{-1}\dfrac{P\sin\alpha}{P + P\cos\alpha}$

$$= \tan^{-1}\frac{P\sin\alpha}{P(1 + \cos\alpha)} = \tan^{-1}\frac{\sin\alpha}{1 + \cos\alpha}$$

$$= \tan^{-1}\frac{2\sin\frac{\alpha}{2}\cos\frac{\alpha}{2}}{2\cos^2\frac{\alpha}{2}} \qquad \left(\because \sin\alpha = 2\sin\frac{\alpha}{2}\cos\frac{\alpha}{2}\right)$$

$$= \tan^{-1} \frac{\sin \frac{\alpha}{2}}{\cos \frac{\alpha}{2}} = \tan^{-1}\left(\tan \frac{\alpha}{2}\right) = \frac{\alpha}{2} \qquad (1.4)$$

It is not essential that one of two forces, should be along x axis. The two forces P and Q may be in any direction (see figure 1.5). The angle between P and Q is 'α' then equation 1.1 gives their resultant. From equation 1.2 the direction of the resultant is obtained. Angle θ will be the angle made by resultant with the direction of force P.

Law of Triangle of Forces. It states that, "if three forces acting at a point be represented in magnitude and direction by the three sides of a triangle, taken in order, they will be in equilibrium."

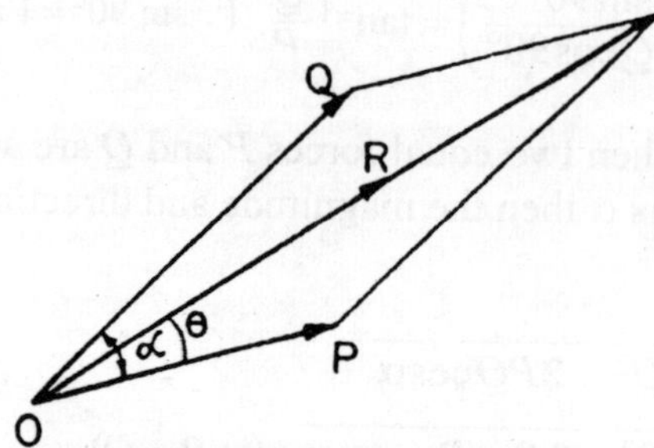

Figure 1.5

Lami's Theorem. It states that, "If there forces acting at a point are in equilibrium, each force will be proportional to the sine of the angle between the other two forces."

Assume the three forces P, Q and R are acting at a point O and they are in equilibrium as shown in Figure 1.6.

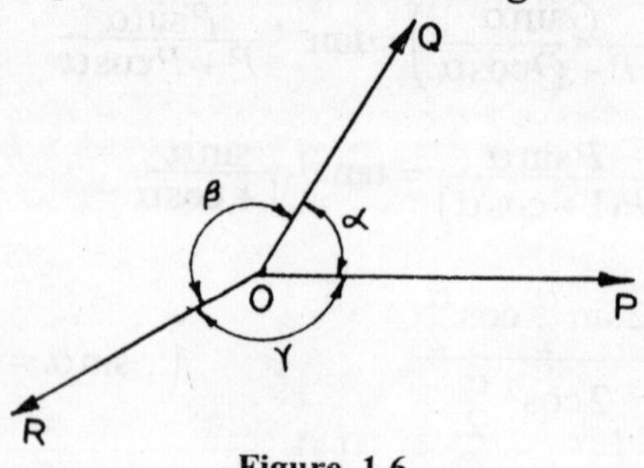

Figure. 1.6

Let α = Angle between force P and Q.

β = Angle between force Q and R.

γ = Angle between force R and P.

Then according to Lami's theorem,

$P \,\alpha$ sine of angle between Q and $R \,\alpha$ sin R.

$$\therefore \qquad \frac{P}{\sin\beta} = \text{constant}$$

Similarly $\dfrac{Q}{\sin\gamma}$ = constant and $\dfrac{R}{\sin\alpha}$ = constant

or
$$\frac{P}{\sin\beta} = \frac{Q}{\sin\gamma} = \frac{R}{\sin\alpha} \qquad (1.5)$$

Proof of Lami's Theorem. The three forces acting on a point, are in equilibrium and hence they can be represented by the three sides of the triangle taken in the same order. Now draw the force triangle as shown in Figure 1.6 (a).

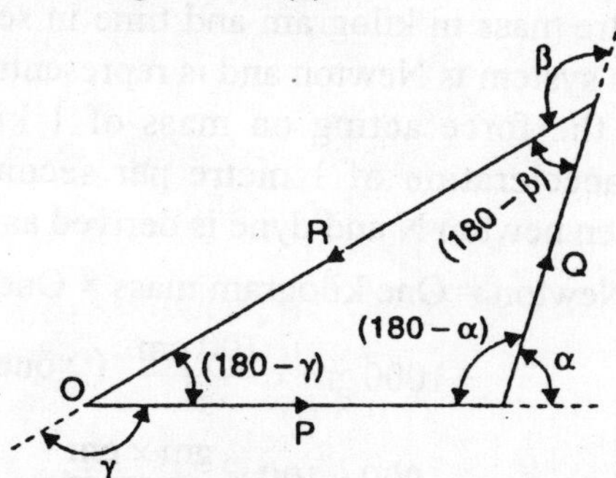

Figure 1.6 (*a*)

Now applying sine rule, we get

$$\frac{P}{\sin(180-\beta)} = \frac{Q}{\sin(180-\gamma)} = \frac{R}{\sin(180-\alpha)}$$

This can also be written

$$\frac{P}{\sin\beta} = \frac{Q}{\sin\gamma} = \frac{R}{\sin\alpha}$$

This is same equation as equation (1.5).

System of Units

The following system of units are usually used

1. C.G.S. (i.e., Centimetre-Gram Second) system of units.

2. M.K.S. (i.e., Metre-Kilogram-Second) system of units.

3. S.I. (i.e., International) system of units.

C.G.S. System of Units. In this system, length is expressed in centimetre, mass in gram and time in second. The unit of force in this system is dyne, which is defined as the force which act on a mass of 1 gram and produce an acceleration of 1 centimetre per second square.

M.K.S. System of Units. In this system, length is expressed in metre, mass in kilogram and time in second. The unit of force in this system is expressed as kilogram force and is represented as kgf.

S.I. System of Units. S.I. is the System International Units; also called the International System of Units. In this system length is express in metre mass in kilogram and time in second. The unit of force in this system is Newton and is represents by N. Newton is defined as the force acting on mass of 1 kilogram which produces an acceleration of 1 metre per second square. The relation between newton N and dyne is derived as—

$$\text{One Newton} = \text{One kilogram mass} \times \text{One metre}$$

$$= 1000\,\text{gm} \times \frac{100\text{ cm}}{\text{s}^2} \quad (\because \text{one kg} = 1000\,\text{gm})$$

$$= 1000 \times 100 \times \frac{\text{gm} \times \text{cm}}{\text{s}^2}$$

$$= 10^5\text{ dyne} \qquad \left\{\because \text{dyne} = \frac{\text{gm} \times \text{cm}}{\text{s}^2}\right\}$$

The units of force like kilo newton and mega newton are used to express the very large magnitude of force. Kilò newton is written as KN.

$$\text{One kilo-newton} = 10^3\text{ newton}$$

or $$1\text{ kN} = 10^3\text{ N}$$

and $$\text{One mega newton} = 10^6\text{ Newton}$$

The large quantities are represented by kilo, mega, giga and terra. They stand for

Kilo = 10^3 and represented by ...... k

Mega = 10^6 and represented by .... M

Giga = 10^9 and represented by ..... G

Tera = 10^{12} and represented by ... T

Thus mega newton means 10^6 newton and is represented by MN. Similarly, giga newton means 10^9 N and is represented by GN. The symbol TN stands for 10^{12} N.

The small quantities are represented by milli, micro, nano and pico. They are equal to

Milli = 10^{-3} and represented by .. m

Micro = 10^{-6} and represented by ... μ

Nano = 10^{-9} and represented by .... n

Pico = 10^{-12} and represented by ... p

Thus milli newton means 10^{-3} newton and is represented by MN. Micro newton means 10^{-6} N and is represented by μN.

The relation between kilogram force (kgf) and newton (N) is given by One kgf = 9.81 N

The force by which the body is attracted towards earth is called its weights. If W = weight of a body, m = mass in kg, then $W = m \times g$ Newtons

If mass, m of the body is 1 kg, then its weight will be,

$$W = 1(\text{kg}) \times 9.81 \frac{\text{m}}{\text{s}^2} = 9.81\text{N}. \qquad \left(\because \text{N} = \text{kg}\frac{\text{m}}{\text{s}^2}\right)$$

Trigonometric Formulae and Expressions

The following are the trigonometric formulae in a right-angled triangle ABC of Figure 1.7.

(*i*) $\sin\theta = \frac{AC}{BC}$

(*ii*) $\cos\theta = \frac{AC}{AB}$

(*iii*) $\tan\theta = \frac{AB}{BC}$

(*iv*) $\sin(A + B) = \sin A \cos B + \cos A \sin B$

(*v*) $\sin(A - B) = \sin A \cos B - \cos A \sin B$

(*vi*) $\cos(A + B) = \cos A \cos B - \sin A \sin B$

(*vii*) $\cos(A - B) = \cos A \cos B + \sin A \sin B$

(*viii*) $\tan(A + B) = \frac{\tan A + \tan B}{1 - \tan A \tan B}$

(*ix*) $\tan(A - B) = \frac{\tan A - \tan B}{1 + \tan A \tan B}$

(*x*) $\sin 2A = 2 \sin A \cos A$

(*xi*) $\sin^2\theta + \cos^2\theta = 1.$

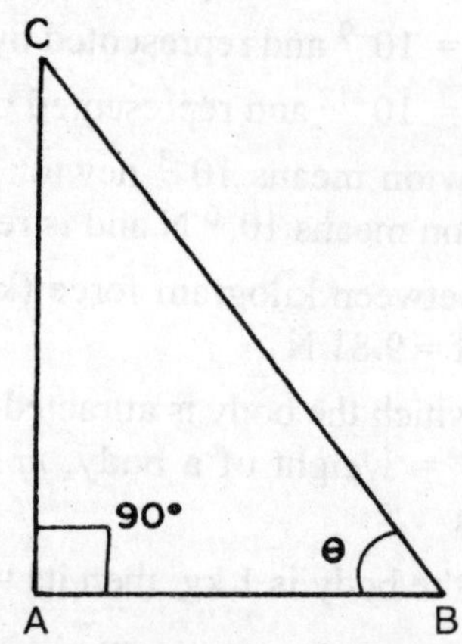

Figure 1.7

Differentiation and Integration

Differentiations

(*i*) Differentiation of a quantity (say A) with respect to x is written as $\frac{d}{dx}(A)$ or $\frac{dA}{dx}$

(*ii*) $\frac{d}{dx}(x^4) = 4x^3$, $\frac{d}{dx}(x^n) = n\,x^{n-1}$ and $\frac{d}{dx}(x) = 1$

(*iii*) $\frac{d}{dx}(8x + 5)^4 = 4(8x + 5)^3 \times 8$

(*iv*) $\frac{d}{dx}(4) = 0$ as differentiation of constant is always zero.

(*v*) $\frac{d}{dx}(u,v) = u \cdot \frac{dv}{dx} + v \cdot \frac{du}{dx}$

[when *u* and v arc functions *of x*]

(vi) Differentiation of trigonometrical functions are given as:

$$\frac{d}{dx}(\sin x) = \cos x$$

$$\frac{d}{dx}(\cos x) = -\sin x$$

$$\frac{d}{dx}(\tan x) = \sec^2 x$$

Resolution of a force

Resolution of a force means "finding its components in two given directions."

Let a given force be *R* making an angle θ with X-axis as shown in Figure 1.8. we have to find the components of the force *R* along *X*-axis and *Y*-axis.

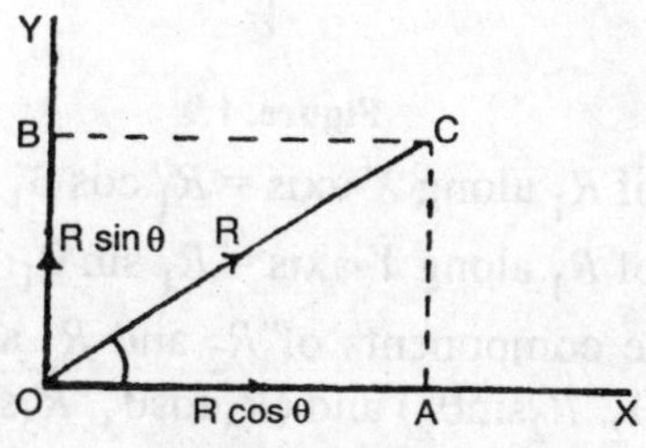

Figure 1.8

Components of *R* along X-axis = *R* cos θ.

Component of *R* along Y-axis = *R* sin θ.

So, the resolution of forces is the process of finding its components in specified directions.

Resolution of a Number of Complanar Forces

The forces acting in one plane are called co-planar forces. Let a number of co-planar forces R_1, R_2, R_3, ... be acting at a point as shown in figure 1.9.

Let θ_1 = Angle made by R_1 with X-axis

θ_2 = Angle made by R_2 with X-axis

θ_3 = Angle made by R_3 with X-axis

H = Resultant component of all forces along X-axis

V = Resultant component of all forces along Y-axis

R = Resultant of all forces

θ = Angle made by resultant with X-axis.

Each force can be resolved into two components, one along X-axis and other along Y-axis.

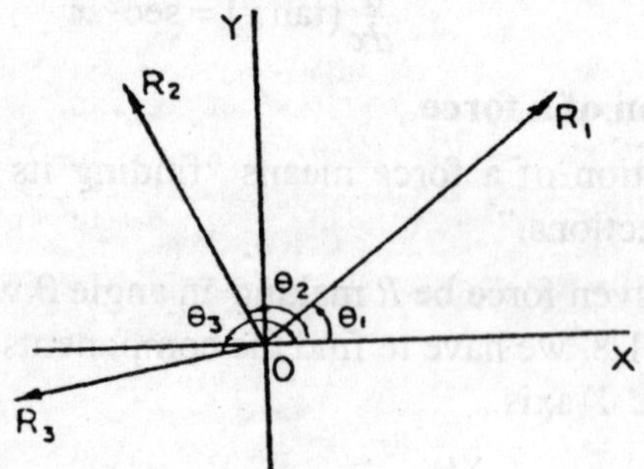

Figure. 1.9

Component of R_1 along X-axis = $R_1 \cos \theta_1$

Component of R_1 along Y-axis = $R_1 \sin \theta_1$.

Similarly, the components of R_2 and R_3 along X-axis and Y-axis are ($R_1 \cos\theta_2$, $R_2 \sin\theta_2$) and ($R_3 \cos\theta_3$, $R_3 \sin\theta_3$) respectively.

Resultant components along X-axis

= Sum of components of all forces along X-axis.

$$\therefore \quad H = R_1 \cos \theta_1 + R_2 \cos \theta_2 + R_3 \cos \theta_3 + ... \quad (1.6)$$

Resultant component along Y-axis.

= Sum of components of all forces along Y-axis.

$$\therefore \quad V = R_1 \sin\theta_1 + R_2 \sin\theta_2 + R_3 \sin\theta_3 + ... \quad (1.7)$$

Then resultant of all the forces, $R = \sqrt{H^2 + V^2}$

The angle made by R with X-axis is given by, $\tan\theta = \frac{V}{H}$ (1.9)

Moment of Force

The moment of the force at any point is defined as the product of a force and the perpendicular distance of the line of action of the force from that point.

P = A force acting on a body as.shown in Figure 1.10.

r = Perpendicular distance between the point O and line of action of the force P.

The moment of the force P about $O = P \times r$

The tendency of the moment P × r is to rotate the body in the clockwise direction about O.

Hence this moment is called clockwise moment. The moment is called anti clock wise moment, if the tandency of rotation is anti clock wise.

Units of Moment

In M.K.S. system the moment is expressed as kgf m whereas in S.I. system, moment is expressed as newton metre (N m).

Effect of Force and Moment on a Body

The linear displacement is caused by the force acting on the body while angular displacement is caused byt he moment. So a body when acted by a numberof co-planar forces will be in equilibrium if:

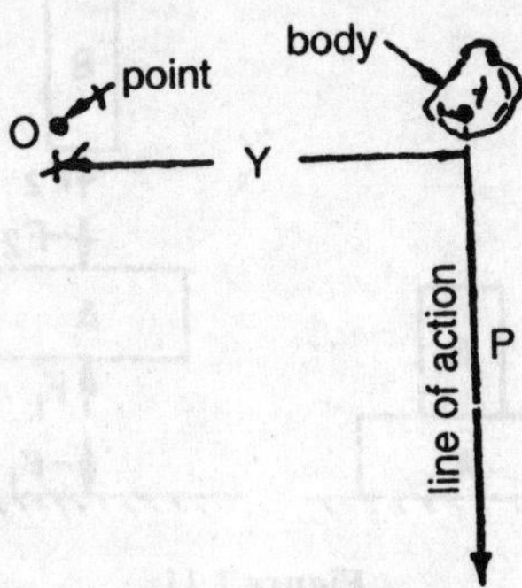

Figure. 1.10

(*i*) Resultant component of forces in the direction of x, y and z are zero *i.e.*, resultant component of forces along any direction is zero.

(*ii*) Resultant moments of forces is zero *i.e.,* clock wise moment is equal to anti-clock wise moments.

Laws of Mechanics

The foundation of mechanics is laid on the following basic laws and principles :

(*i*) Newton's first and second laws of motion

(*ii*) Newton's third law

(*iii*) The gravitational law of attraction

(*iv*) The parallelogram law

(*v*) The Principle of Transmissibility of forces.

Newton's First and Second Laws of Motion

Newton's first law states, "Every body continues in a state of rest or uniform motion in a straight line unless it is compelled to change that state by some external force acting on it."

Newton's second law states, "The net external force acting on a body in a direction is directly proportional to the rate of change of momentum in that direction."

Newton's Third Law

Newton's third law states, "To every action there is always equal and opposite reaction."

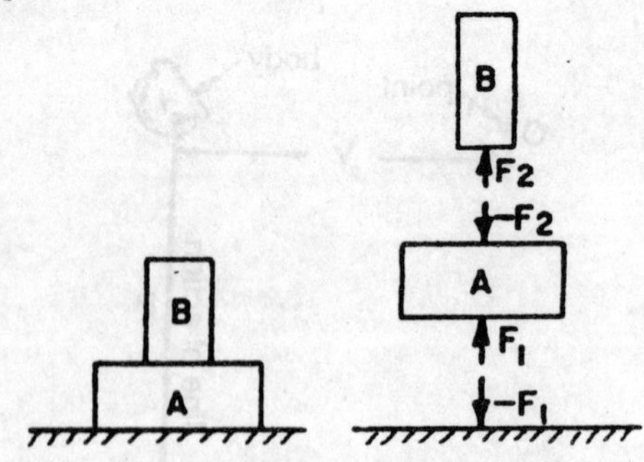

Figure 1.11

Figure 1.11 shown two bodies A and B which are placed one above the other on a horizontal surface.

Here F_1 = Force exerted by horizontal surface on body A

(action)

$-F_1$ = Force exerted by body *A* on horizontal surface (reaction)

F_2 = Force exerted by body *A* on body *B* (action)

$-F_2$ = Force exerted by body *B* on body *A* (reaction)

The Gravitational Law of Attraction. It states that two bodies are attracted towards each other along the line connecting them with a force which is directly proportional to.the product of their masses and inversely proportional to the square of the distance between their centres.

Refer to Figure 1.12 (a).

Let m_1 = Mass of first body

m_2 = Mass of second body

r = Distance between the centre of bodies

F = Force of attraction between the bodies.

Then according to the law of gravitational attraction.

$F \propto m_1 \cdot m_2$

or
$$F \propto \frac{m_1 m_2}{r^2}$$

or
$$F = G\frac{m_1 m_2}{r^2}$$

where G = Universal gravitational constant of proportionality.

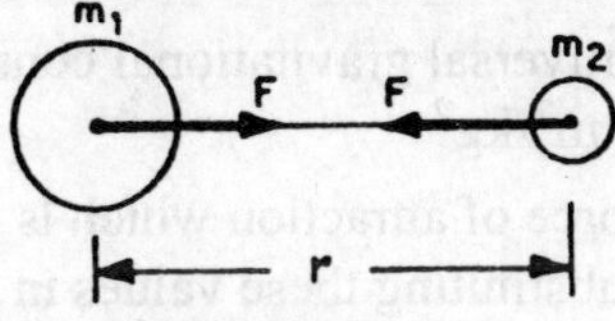

Figure 1.12 ***(a)***

In equation (1.10), F is expressed in N, m_1 and m_2 in kg and r in m. Hence dimensionally, equation (1.10) can be written as

$$N = G\frac{\text{kg} \times \text{kg}}{\text{m}^2} \quad \text{or} \quad G = \frac{\text{N m}^2}{\text{kg}^2} \qquad (i)$$

But $\quad N = 1\text{kg} \times \frac{1\text{m}}{\text{s}^2} \quad$ or $\quad N = \text{kg} \times \frac{\text{m}}{\text{s}^2}$

Substituting the value of N is equation (*i*),

$$G = \left(\text{kg} \times \frac{\text{m}}{\text{s}^2}\right) \times \frac{\text{m}^2}{\text{kg}^2} = \frac{\text{m}^2}{\text{kg s}^2} \qquad (ii)$$

Hence from equations (*i*) and (*ii*), dimension of Universal Gravitational Constant G is N m²/kg² or m³/kg s².

The value of G is 6.67×10^{-11} N m²/kg² or m³/kg s².

In equation (1.10), if m_1=1 kg, m_2 =1 kg and r= I m, then F = G. This means that the force of attraction between two bodies of mass 1 kg each when they are at a distance of 1 m apart, will be 6.67×10^{-11} N i.e., 0.0000000000667 N. This force is very very small.

Weight

The weight of a body is defined with the help of law of gravitation. The force with which a body is attracted towards the centre of earth is known as its weight.

Let $\quad M$ = Mass of the body

M_E = Mass of the earth = 5.9761×10^{24} kg

r = Distance between the centres of the earth and the body

= 6.371×10^6 m (*i.e.*, radius of earth)

G = Universal gravitational constant = 6.67×10^{-11} N m²/kg²

F = Force of attraction which is equal to weight (W)

Substituting these values in equation (1.10), we get

$$W = G\frac{M_E \times M}{r^2} \qquad (\because F = W,\ m_1 = M,\ m_2 = M_E)$$

$$= \frac{6.67 \times 10^{-11}\left(\frac{\text{Nm}^2}{\text{kg}^2}\right) \times 5.9761 \times 10^{24}(\text{kg}) \times M(\text{kg})}{(6.371 \times 10^6\,\text{m})^2}$$

$$= \frac{6.67 \times 10^{-11} \times 5.9761 \times 10^{24} M}{6.731^2 \times 10^{12} \times \text{m}^2} \frac{\text{N m}^2}{\text{kg}^2}$$

$$\times \text{kg} \times \text{kg} = (9.81 \times M)\ \text{N}$$

where 9.81 is acceleration due to gravity and is denoted by 'g'.

$\therefore\ W = \text{g} \times M$ or $M \times g$

Actually the term $\frac{GM_E}{r^2}$ is equal to $9.81 \frac{\text{m}}{\text{s}^2}$, which is represented by 'g'.

The Parallelogram Law

It states that if two forces acting at a point are represented in magnitude and direction by the two adjacent sides of a parallelogram, then their resultant is represented in magnitude and direction by the diagonal of that parallelogram passing through that point.

The Principle of Transmissibility of Forces

According to this principal if a force, acting at a point on a rigid body, is shifted to any other point which is on the line of action of the force, the external effect of the force on the body remains unchanged.

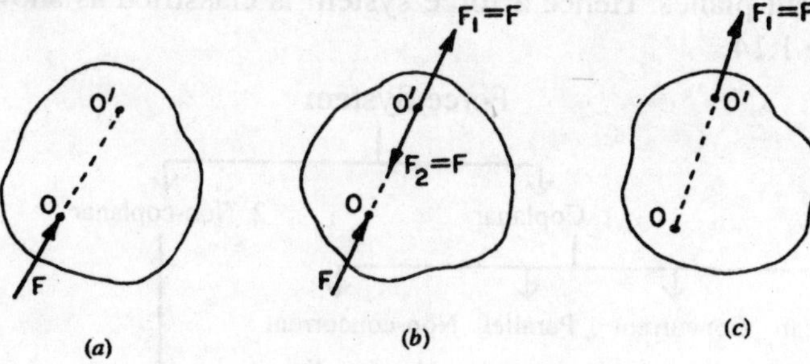

Figure 1.13

For example, let us consider a force F acting at point O on a rigid body as shown in figure 1.13. (a). On this rigid body, "there is another point O' in the line of action of the force F. Suppose at this point O', two equal and opposite forces F, and F_2 (each equal to F and collinear with F) are applied as shown in figure 1.13 (b).

The force F and F_2, being equal and opposite, will cancell each other, leaving a force F, at point O' as shown in figure 1.13 (c). But force F_1 is equal to force F.

The original force F acting at point O, has been transferred to point O' which is along the line of action of F without changing the effect of the force on the rigid body. Hence any force acting at a point on a rigid body can be transmitted to act at any other point along its line of action without changing its "effect on the rigid body. This proves the principle of transmissibility of a force.

COPLANAR COLLINEAR AND CONCURRENT FORCES

The forces in a plane are known as co-planar forces. The forces which are having common lines of actions are called as collinear while the forces which intersect at a common point are said to be concurrent. When several forces act on a body, then they are called *a* force system or a system of forces. The system in which all the forces lie in the same plane is known as *coplanar force system.*

Classification of a Force System

A force system may be coplanar or non-coplanar. If in a system all the forces lie in the same plane then the force system is known as coplanar. But in non-coplanar system all the forces lie in different planes. Hence a force system is classified as shown in figure 1.14.

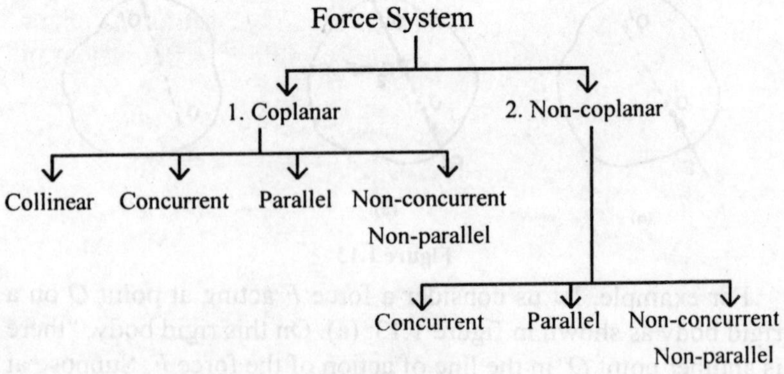

Figure 1.14

In this chapter,, we shall discuss only coplanar force system, in it the forces may be

(*i*) Collinear

(*ii*) Concurrent

(*iii*) Parallel

(*iv*) Non-concurrent, non-parallel (or General system of forces)

Coplanar Collinear

Three forces F_1, F_2 and F_3 acting in a plane are shown in figure 1.15. These three forces are in the same line *i.e.*,these three forces are having a common line of action. This system of forces is known as coplanar collinear force system. Hence in coplanar

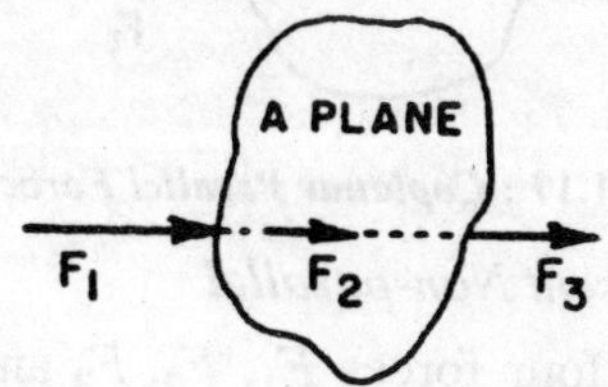

Figure 1.15 : *Coplanar Collinear Forces.*

collinear system of forces, all the forces act in the same plane and have a common line of action.

Coplanar Concurrent

Figure 1.16 shows three forces F_1, F_2 and F_3 acting in a plane and these forces intersect at a common point *O*. This system of

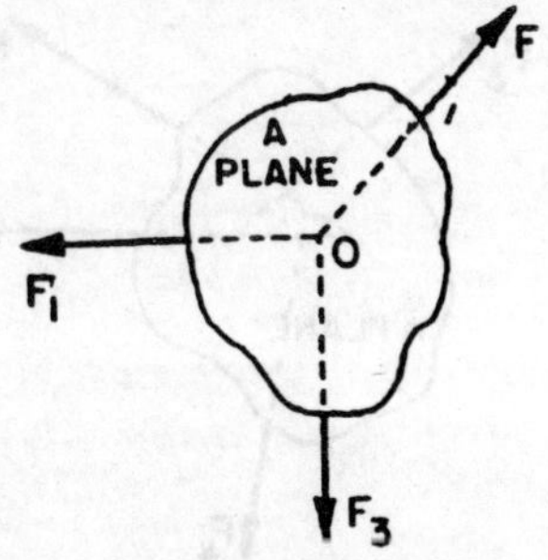

Figure 1.16 : *Concurrent Coplanar Forces.*

forces is known as coplanar concurrent force system. Thus it can be said that in coplanar concurrent system of forces, all the forces act in the same plane and they intersect at a common point.

Coplanar Parallel

Figure 1.17 shows three forces F_1, F_2 and F_3 acting in a plane and these forces are parallel to one another. This system of forces is known as coplanar parallel system of force system. Hence in coplanar parallel system of forces, all the forces act in the same plane and are parallel to each other.

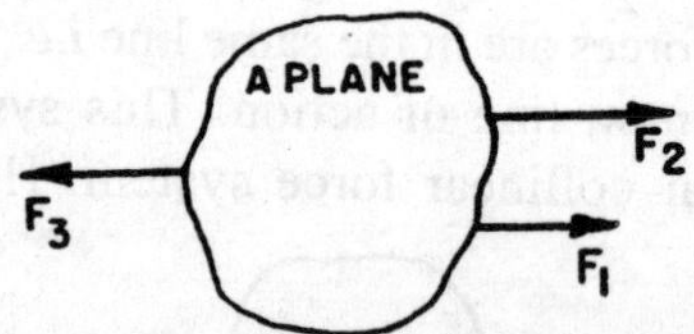

Figure 1.17 : *Coplanar Parallel Forces*

Coplanar Non-concurrent Non-parallel

Figure 1.18 shows four forces F_1, F_2, F_3 and F_4 acting in a plane. The lines of action of these forces lie in the same plane but they are neither parallel nor intersect at a common point. This system of forces is known as coplanar non-concurrent non-parallel force system. Hence in coplanar non-concurrent non-parallel system of forces, all the forces act in the same plane. But the forces are neither parallel nor meet at a common point. This force system is also known as *general system of forces.*

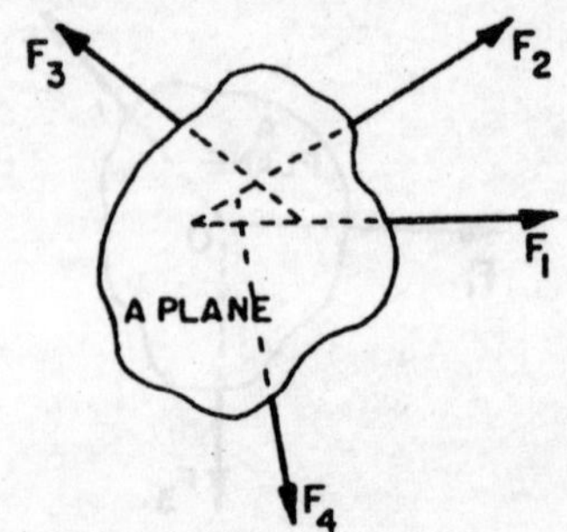

Figure 1.18 : *Non-concurrent Non-parallel.*

Resultant of Several Forces

When a single force can replace a number of force acting on a rigid body, without causing any changein the external effect on the body, it is known as resultant force.

When a no. of co-planar forces are acting on a rigid body, and these can be replaced by a single force of same intensity then this single force is called as the resultant of several forces.

Resultant of Coplanar Forces

The resultant of coplanar forces may be obtained by following two methods

1. Graphical method
2. Analytical method.

The resultant of the following coplanar forces will be determined by the above two methods

(*i*) Resultant of collinear coplanar forces

(*ii*) Resultant of concurrent coplanar forces.

Resultant of Collinear Coplanar Forces

Collinear coplanar forces are those forces which act ir, the same plane and have a common line of action. The resultant of these forces can be obtained by analytical method or graphical method.

Analytical method

If all the forces are acting in the same direction than the resultant is obtained by adding all the forces. If any one of the forces is acting in opposite direction than the resultant is obtained by substracting that particular forces.

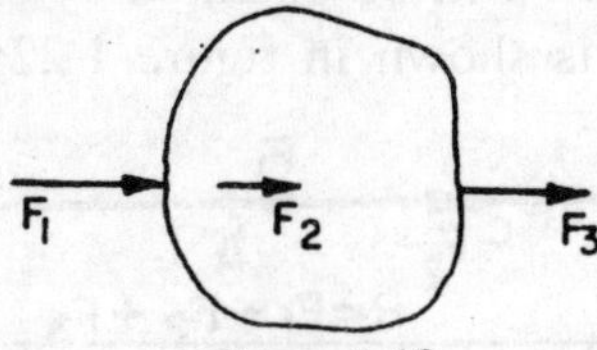

Figure 1.19

Figure 1.19 shows three collinear coplanar forces F_1, F_2 and F_3 acting on a rigid body in the same direction. Their resultant R will be sum of these forces.

$$\therefore \qquad R = F_1 + F_2 + F_3 \qquad (1.11)$$

If anyone of these forces (say force F_2) is acting in the opposite direction, as shown in Figure 1.20, then their resultant will be given by

$$R = F_1 - F_2 + F_3 \qquad (1.12)$$

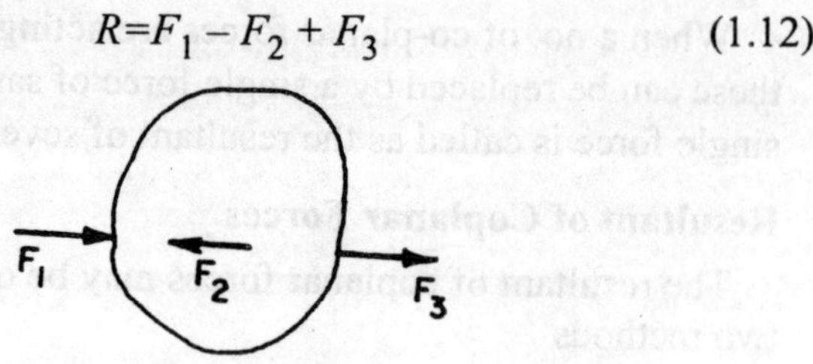

Figure 1.20

Graphical Method

First of all, we choose a suitable scale than vectors ar drawn to the chosen scale amd than these vectors are either added or substracted to find the resultant. The resultant of the three collinear forces F_1, F_2 and F_3 acting in the same direction will be obtained by adding all the vectors. In figure 1.21 the force $F_1 = ab$ to some scale, force $F_2 = bc$ and force $F_3 = cd$. Then the length ad represents the magnitude of the resultant on the scale chosen.

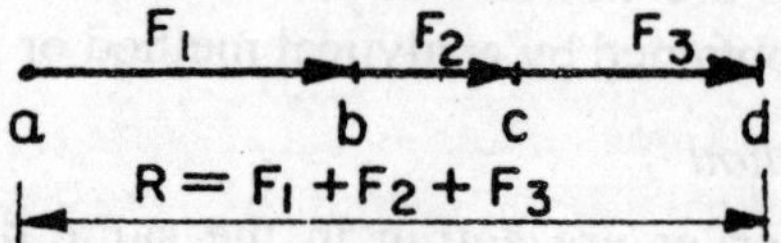

Figure 1.21

The resultant of the forces F_1, F_2 and F_3 acting on a body shown in figure 1.20 will be obtained by subtracting the vector F_2. This resultant is shown in figure 1.22; In which the force

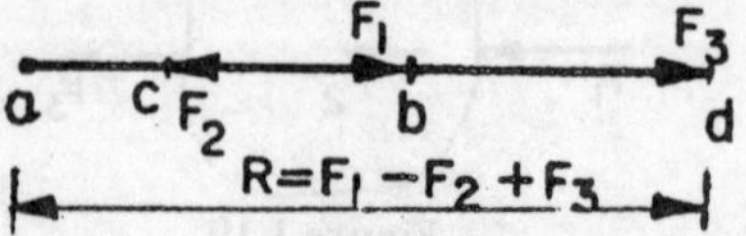

Figure 1.22

$F_1 = ab$ to some suitable scale. This force is acting from a to b. The force F_2 is taken equal to be on the same scale in opposite direction. This force is acting from *b* to *c*. The force F_3 is taken equal to *cd*. This force is acting front *c* to *d*. The resultant force is represented in magnitude by *ad* on the chosen scale.

Resultant of Concurrent Coplanar Forces

The forces which act in the same plane and intersect at a common point are called concurrent coplanar forces. We will study the following two cases:

(*i*) When two forces act at a point

(*ii*) When more than two forces act at a point.

When two forces act at a point

(*a*) *Analytical Method*

When two forces act at a point, their resultant is determined the law of parallelogram of forces. The magnitude of resultant is obtained from equation (1.1) and the direction of resultant with one of the forces is obtained from equation (1.2).

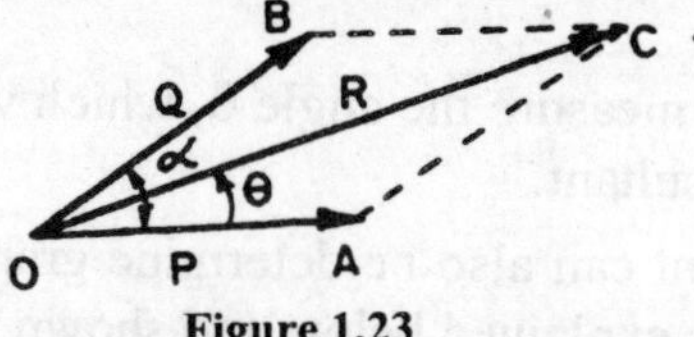

Figure 1.23

Suppose two forces *P* and *Q* act at point *O* as shown in figure 1.23 and α is the angle between them. Let θ is the angle made by the resultant R with the direction of force P.

Forces *P* and *Q* form two sides of a parallelogram and according to the law, the diagonal through the point *O* gives the resultant *R* as shown.

The magnitude of resultant is given by

$$R = \sqrt{P^2 + Q^2 + 2PQ\cos\alpha}$$

The above method of determining the resultant is also known as the *cosine law method*. The direction of the resultant with the force *P* is given by

$$\theta = \tan^{-1}\left(\frac{Q\sin\alpha}{P + Q\cos\alpha}\right)$$

(b) Graphical Method

(*i*) Choose a convenient scale to represent the forces *P* and Q.

(*ii*) From point *O*, draw a vector *Oa* = *P*.

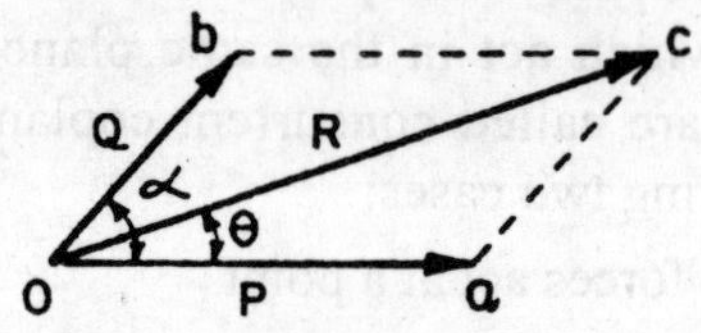

Figure : 1.24

(*iii*) Now from point *O*, draw another vector *Ob* = *Q* and at an angle of α as shown in figure 1.24.

(*iv*) Complete the parallelogram by drawing lines ac || to *Ob* and and *be* || to *Oa*.

(*v*) Measure the length *OC*.

Then resultant *R* will be equal to length *OC* × chosen scale.

(*vi*) Also measure the angle θ, which will give the direction of resultant.

The resultant can also be determine graphically by drawing a triangle *oac* as explained below and shown in figure 1.25.

(*i*) Draw a line.oa parallel to *P* and equal to *P*.

(*ii*) From *a*, draw a vector *ac* at an angle a with the horizontal and cut *ac* equal to *Q*.

(*iii*) Join *oc*. Then *oc* represents the magnitude and direction of resultant R.

Magnitude of resultant *R* = Length *OC* × chosen scale. The direction of resultant is given by angle θ. Hence measure the angle θ.

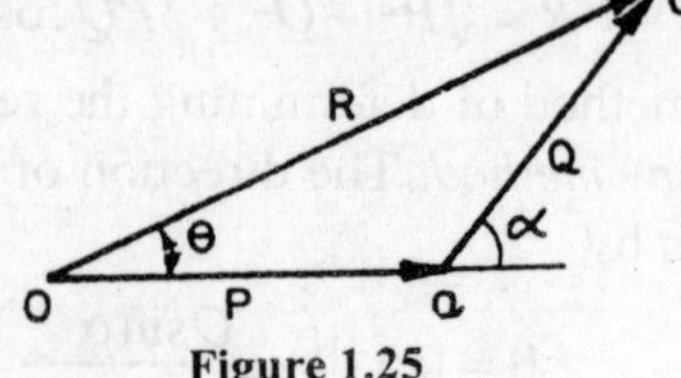

Figure 1.25

When more than two forces act at a point

(a) Analytical Method

Analytically, thye resultant of three of more forces acting at a point is determined by rectangular component method. In this method all the forces acting at a point are firstly resolved into vertical and horizontal components and then their algebric summation is determined seperately. The summation of horizontal component is represented as ΣH and of vertical component as ΣV. The resultant R is given by

$$R = \sqrt{(\Sigma H)^2 + (\Sigma V)^2}$$

The angle made by the resultant with horizontal is given by

$$\tan\theta = \frac{(\Sigma V)}{(\Sigma H)}$$

∴ Let four forces F_1, F_2, F_3 and F_4 act at a point O as shown in figure 1.26.

The inclination of the forces is indicated with respect to horizontal direction. Let

θ_1 = Inclination of force F_1 with OX

θ_2 = Inclination of force F_2 with OX'

θ_3 = Inclination of force F_3 with OX'

θ_4 = Inclination of force F_4 with OX.

The force F_1 is resolved into horizontal and vertical components and these components are shown in figure 1.26 (a). Similarly, figures 1.26 (*b*), (*c*) and (*d*) shows the horizontal and vertical components of forces F_2, F_3 and F_4 respectively. The various horizontal components are:

$$F_1 \cos\theta_1 \rightarrow (+)$$

$$F_2 \cos\theta_2 \leftarrow (-)$$

$$F_3 \cos\theta_3 \leftarrow (-)$$

$$F_4 \cos\theta_4 \rightarrow (+)$$

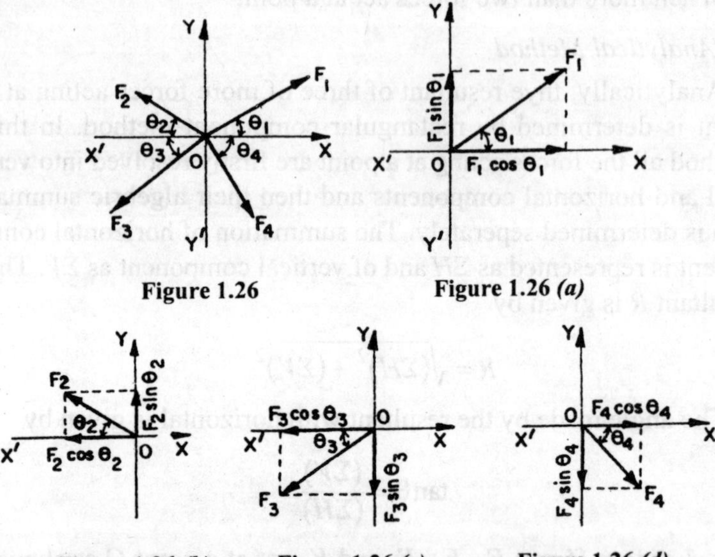

Figure 1.26 Figure 1.26 *(a)*

Figure 1.26 *(b)* Figure 1.26 *(c)* Figure 1.26 *(d)*

∴ Summation or algebraic sum of horizontal components:

$$\Sigma H = F_1 \cos\theta_1 - F_2 \cos\theta_2 - F_3 \cos\theta_3 + F_4 \cos\theta_4$$

Similarly, various vertical components of all forces are:

$$F_1 \sin\theta_1 \uparrow (+)$$

$$F_2 \sin\theta_2 \uparrow (+)$$

$$F_3 \sin\theta_3 \downarrow (-)$$

$$F_4 \sin\theta_4 \downarrow (-)$$

∴ Summation or algebraic sum of vertical components:

$$\Sigma V = F_1 \sin\theta_1 + F_2 \sin\theta_2 - F_3 \sin\theta_3 - F_4 \sin\theta_4$$

Then the resultant will be given by $R = \sqrt{(\Sigma H)^2 + (\Sigma V)^2}$ 1.11

And the angle (θ) made by resultant with x-axis is given by $\tan\theta = \dfrac{(\Sigma V)}{(\Sigma H)}$ 1.12

(b) Graphical method

Graphically, the resultant of sexual forces acting at a point is obtained by the help of polygon law of forces which states

"If a number of coplanar forces are acting at a point such as they can be represented in magnitude and direction by the sides of a polygon taken in the same order, then their resultant is repr sented in magnitude and direction by the closing side of the polygon taken in the opposite order.

Let the four forces F_1, F_2, F_3 and F_4 act at a point O as shown in figure 1.27. The resultant is obtained graphically by drawing polygon of forces as explained below and shown in figure 1.27 *(a)*.

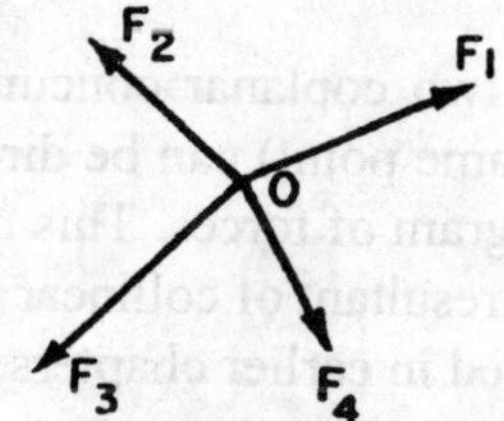

Figure 1.27

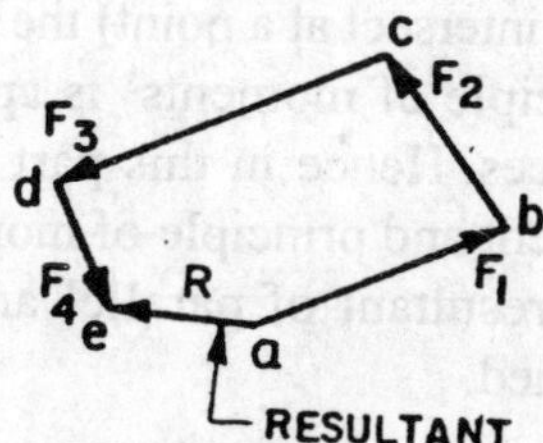

Figure 1.27 *(a)*

(i) Choose a suitable scale to represent the given forces.

(ii) Take any point *a*. From *a*, draw vector *ab* parallel to OF_1. Cut *ab* = force *F*, to the scale.

(iii) From point *b*, draw *be* parallel to OF_2. Cut *be* = force F_2.

(iv) From point C, draw *cd* parallel to OF_3. Cut *cd* = force F_3.

(v) From point *d*, draw *de* parallel to OF_4. Cut *de* = force F_4.

(*vi*) Join point a to e. This is the closing side of the polygon. Hence ae represents the resultant in magnitude and direction.

Magnitude of resultant R = Length $ae \times$ scale.

The resultant is acting from a to e.

COPLANAR PARALLEL FORCES AND VARIGNON'S THEOREM

Parallel forces are the forces having their line of action parallel to one another. The two parallel forces will never meet or intersect at a point.

The resultant of two coplanar concurrent forces (i.e., forces intersecting at the same point) can be directly determined by the method of parallelogram of forces. This method along with other methods for finding resultant of collinear and concurrent coplanar forces, were discussed in earlier chapters.

The parallel forces have their lines of action parallel to each other. So, for finding the resultant of two parallel forces, (two parallel forces do not intersect at a point) the parallelogram cannot be drawn. The 'Principle of moments' is applied to find out the resultant of such forces. Hence in this part first we will discuss the concepts of moment and principle of moments. Thereafter the methods of finding resultant of parallel and even non-parallel forces will be explained.

Moment of a Force

The product of a force and the perpendicular distance of the line of action of the force from that point is known as moment of the force about that point.

Let F = A force acting on a body as shown in figure 1.28

r = Perpendicular distance from the point O on the line of.action of force F.

Then moment (M) of the force F about O is given by, $M = F \times r$

If the tandency of the moment is to rotate the body in the clockwise direction about O than this moment is called clockwise moment. If the tendency of the moment is to rotate the body in anti-clock wise direction then the moment is known as anti-clock

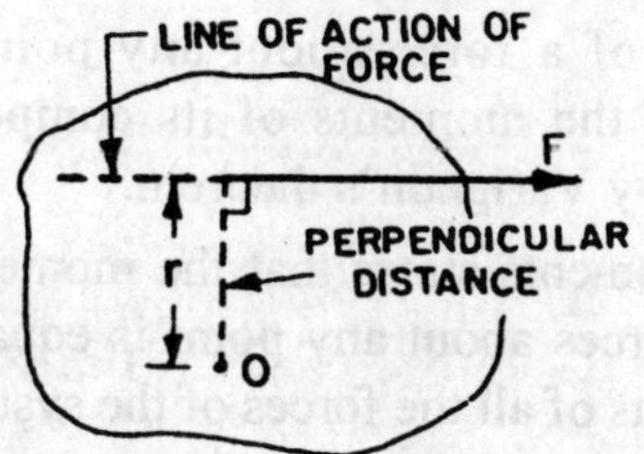

Figure 1.28

wise moment. If clockwise moment is taken –ve then anti-clockwise moment will be + ve.

In S.I. system, moment is expressed in N m (Newton metre).

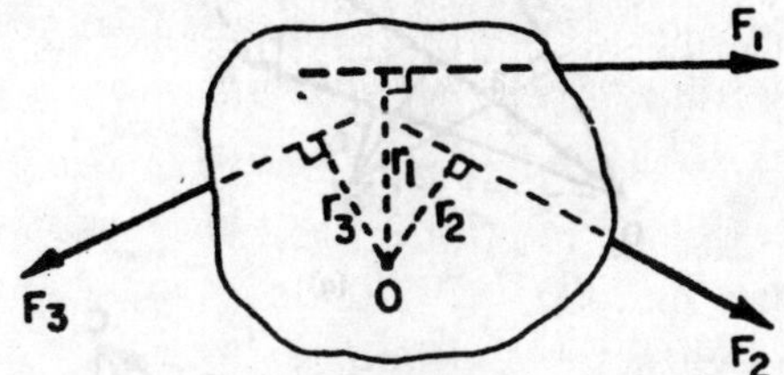

Figure 1.29

Figure 1.29 shows a body on which three forces F_1, F_2 and F_3 are acting. If it is required to find resultant moments if these forces about point *O*.

Let r_1 = Perpendicular distance from *O* on the line of action of force F_1.

r_2 and r_3 = Perpendicular distances from *O* on the lines of action of force F_2 and F_3 respectively.

Moment of F_1 about $O = F_1 \times r_1$ (clockwise) (–)

Moment of F_2 about $O = F_2 \times r_2$ (clockwise) (–)

Moment of F_3 about $O = F_3 \times r_3$ (anti-clockwise) (+)

The resultant moment will be *algebraic sum* of all the moments.

∴ The resultant moment of F_1, F_2 and F_3 about *O*

$$= -F_1 \times r_1 - F_2 \times r_2 + F_3 \times r_3$$

Varignon's Theorem (or Principle of Moments)

"The moment of a force about any point is equal to the *algebraic sum* of the moments of its components about *that point.*" as stated by Varignon's theorem.

Principle of moments states that the moment of the resultant of a number of forces about any point is equal to the *algebraic sum* of the moments of all the forces of the system about the same point.

Proof of Varignon's Theorem

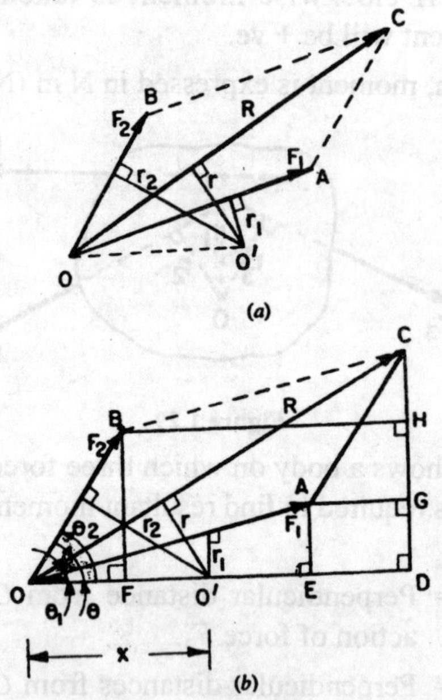

Figure 1.30

Figure 1.30 (a) shows two forces F_1 and F_2 acting at point O. These forces are represented in magnitude And direction by OA and OB. OC which is the diagonal of parallelogram $OACB$ represent the resultant R in magnitude and direction. Let O is the point in the plane about which moments of F_1, F_2 and Rare to be determined. From point O, draw perpendiculars on OA, OC and

OB.

Let r_1 = Perpendicular distance between F_1 and O'.

r = Perpendicular distance between R and O'.

r_2 = Perpendicular distance between F_2 and O'.

Then according to Varignon's principle;

Moment of R about O' must be equal to algebraic sum of moments of F_1 and F_2 about O'.

or $$R \times r = F_1 \times r_1 + F_2 \times r_2$$

Now refer to figure 1.30 (b) join *OO′* and produce it to *D*. From points *C, A* and *B* draw perpendiculars on *OD* meeting at *D, E* and respectively. From A and *B* also draw perpendiculars on *CD* meeting the line *CD* at G and *H* respectively.

Let θ_1 = Angle made by F_1 with *OD*,

θ = Angle made by R with *OD*, and

θ_2 = Angle made by F_2 with *OD*.

In figure 1.30 (b), *OA* = *BC* and also *OA* parallel to *BC*, hence the projection of *OA* and *BC* on the same vertical line *CD* will be equal i.e., *GD* = *CH* as *GD* is the projection of *OA* on *CD* and *CH* is the projection of *BC* on *CD*.

Then from figure 1.30 (b), we have

$F_1 \sin \theta_1 = AE = GD = CH$

$F_1 \cos \theta_1 = OE$

$F_2 \sin \theta_2 = BF = HD$

$F_2 \cos \theta_2 = OF = ED$ $\because$ (*OB* = *AC* and also *OB* || *AC*. Hence projections of *OB* and *AC* on the same horizontal line *OD* will be equal i.e., *OF* = *ED*)

$R \sin \theta = CD$

$R \cos \theta = OD$

Let the length $OO' = x$.

Then $x \sin \theta_1 = r_1$, $x \sin \theta = r$ and $x \sin \theta_2 = r_2$

Now moment of R about O'

$= R \times (\theta \text{ distance between } O' \text{ and } R) = R \times r$

$= R \times x \sin\theta$ ($r = x\sin\theta$)

$= (R\sin\theta) \times x$

$= CD \times x$ ($R\sin\theta = CD$)

$= (CH + HD) \times x$

$= (F_1 \sin\theta_1 + F_2 \sin\theta_2) \times x$

($CH = F_1 \sin\theta_1$ and $HD = F_2 \sin\theta_2$)

$= F_1 \times x \sin\theta_1 + F_2 \times x \sin\theta_2$

$= F_1 \times r_1 + F_2 \times r_2$

($x\sin\theta_1 = r_1$ and $x\sin\theta_2 = r_2$)

= Moment of F_1 about O' + Moment of F_2 about O'.

Hence moment of R about any point in the algebraic sum of moments of its components (i.e., F, and F_2) about a same point. Hence Varignon's principle is proved.

The principle of moments (or Varignon's principle) is not restricted to only two concurrent forces but it is also applicable to any coplanar force system, i.e., concurrent or non-concurrent or parallel force system.

Types of Parallel Forces

The two important types of parallel forces are mentioned below

1. Like parallel forces,
2. Unlike parallel forces.

Like parallel forces

Like parallel forces are the forces which are acting in the same direction. In figure 1.31 two parallel forces F_1 and F_2 are shown. They are acting in the same direction. Hence they are called as like parallel forces. Like parallel forces may be equal or unequal in magnitude.

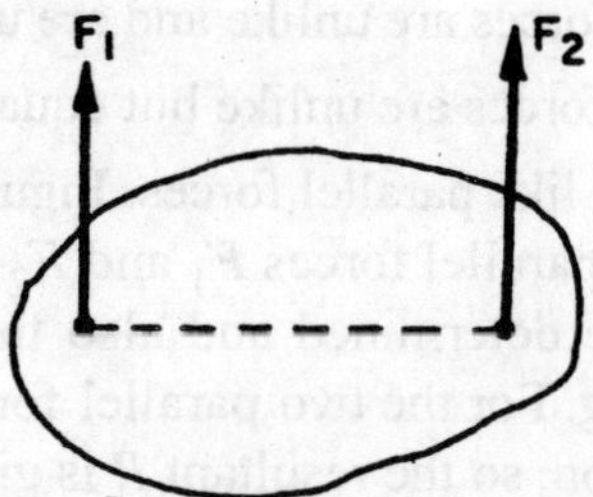

Figure 1.31

Unlike Parallel Forces

The parallel forces which act in the opposite direction are called as unlike parallel forces. In figure 1.32, two parallel forces F_1, F_2 are acting in opposite direction. Hence they are called as unlike parallel forces. These forces may be equal or unequal in magnitude.

The unlike parallel forces are divided into (*i*) unlike equal parallel forces, and (*ii*) unlike unequal parallel forces.

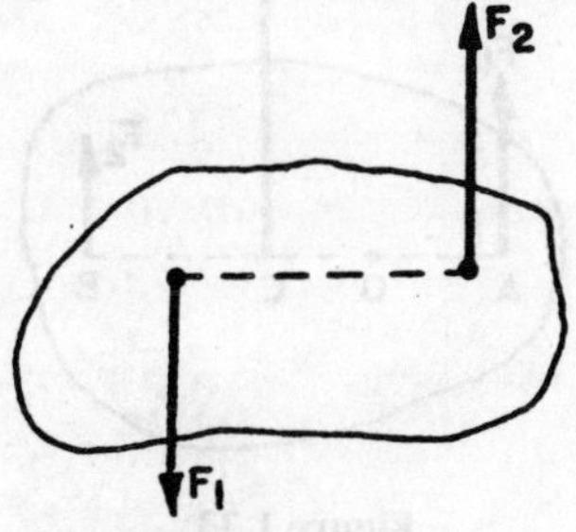

Figure 1.32

The forces which are equal in magnitude but are acting in opposite directions are known as unlike equal parallel forces.

The forces which act in opposite direction but are unequal in magnitude, are called unlike unequal parallel forces.

Resultant of Two Parallel Forces

The resultant of following two parallel forces will be considered

1. Two parallel forces are like.

2. Two parallel forces are unlike and are unequal in magnitude.
3. Two parallel forces are unlike but equal in magnitude.

Resultant of two like parallel forces. Figure 1.33 shows a body on which two like parallel forces F_1 and F_2 are acting. Resultant R is required to be determined and also the point at which the resultant R is acting. For the two parallel forces which are acting in the same direction, so the resultant R is given by,

$$R = F_1 + F_2$$

Varignon's principle is used in order to find the point at which the resultant is acting. According to this, the algebraic sum of moments of F_1 and F_2 about any point should be equal to the moment of the resultant (R) about that point. Now arbitrarily choose any, point O along line AB and take moments of all forces about this point.

Moment of F_1 about $O = F_1 \times AO$ (clockwise) (–)

Moment of F_2 about $O = F_2 \times BO$ (anti-clockwise) (+ve)

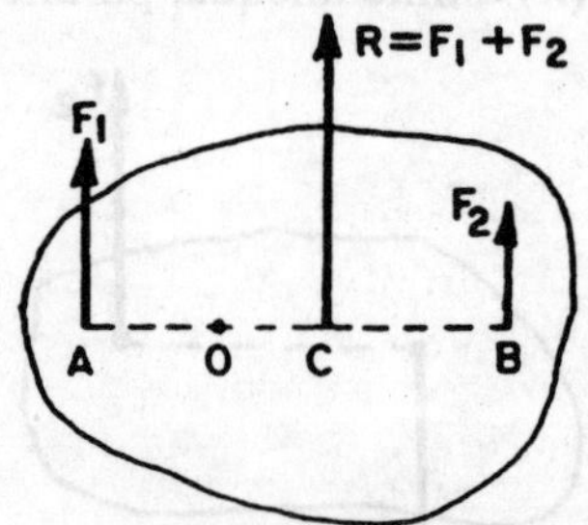

Figure 1.33

∴ Algebraic sum of moments of F_1 and F_2 about O

$$= F_1 \times AO + F_2 \times BO$$

Moment of resultant about $O = R \times OC$ (anti-clockwise)(+)

But according to principle of moments the algebraic sum of moments of F_1 and F_2 about O should be equal to the moment of resultant about the same point O.

$$\therefore -F_1 \times AO + F_2 \times BO = +R \times CO = (F_1 + F_2) \times CO$$

$$(\quad R = F_1 + F_2)$$

or $$F_1 (AO + CO) = F_2(BO - CO)$$

or $$F_1 \times AC = F_2 \times BC$$

$$(\; AO + CO = AC \text{ and } BO\text{-}CO = BC)$$

or $$\frac{F_1}{F_2} = \frac{BC}{AC}$$

The above relation shows that the resultant R acts at the point C, parallel to the lines of action of the given forces F, and F_2 in such a way that the resultant divides the distance AB in the ratio inversely proportional to the magnitudes of F, and F_2. Also the point C lies in line AB *i.e.,* point C is not outside AB.

The location of the point C, at which the resultant R is acting, can also be determined by taking moments about points A of Fig. 1.33. As the force F, is passing through A, the moment of F, about A will be zero.

The moment of F_2 about $A = F_2 \times AB$ (anti-clockwise) (+)

Algebraic sum of moments of F_1 and F_2 about O

$$= O + F_2 \times AB = F_2 \times AB \text{ (anti-clockwise) (+)} \qquad (i)$$

The moment of resultant R about A

$$= R \times AC \text{ (anti-clockwise)(+)} \qquad (ii)$$

But according to the principle of moments, the algebraic sum of moments of F_1 and F_2 about A should be equal to the moment of resultant about the same point A. Hence equating equations (*i*) and (*ii*),

$$F_2 \times AB = R \times AC$$

But $R = (F_1 + F_2)$ hence the distance AC should be less than AB. Or in other words, the point C will lie inside AB.

Resultant of Two Unlike Parallel Forces : (Unequal in magnitude). Two unlike parallel forces F_1 and F_2 are acting on a body as shown in the figure 1.34. The two forces are unequal in magnitude. Let us assume that force F_1 is more than F_2. Now, we have to determine resultant R and also the point at which the resultant R *is* acting. For the two parallel forces, which are acting in opposite direction, obviously the resultant is given by,

$$R = F_1 - F_2$$

Let the resultant R is acting at C as shown in Figure 1.34. In order to find the point; C, at which the resultant is acting, principle of moments is used. Choose arbitrarily any point O in line AB. Take the moments of all forces (i.e., F_1, F_2 and R) about this point.

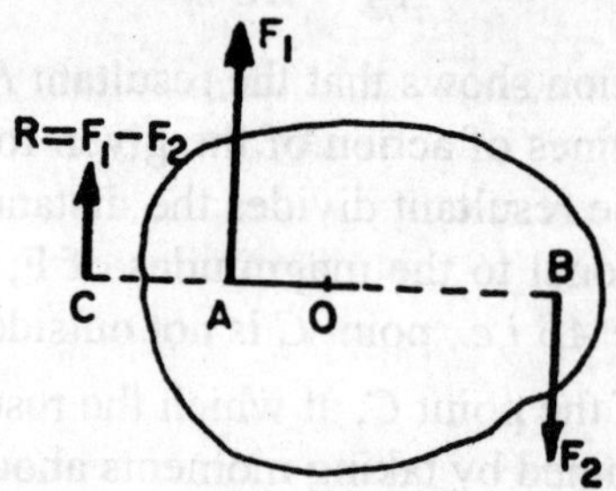

Figure 1.34

Moment of F_1 about $O = F_1 \times AO$ (clockwise)

Moment of F_2 about $O = F_2 \times BO$ (clockwise)

Algebraic sum of moments of F_1 and F_2 about O

$$= F_1 \times AO + F_2 \times BO \qquad (i)$$

Moment of resultant force R about O

$$= R \times CO \text{ (clockwise)}$$

$$= (F_1 - F_2) \times CO \qquad (\because R = F_1 - F_2)$$

$$= F_1 \times CO - F_2 \times CO \qquad (ii)$$

But according to the principle of moments, the algebraic sum of moments of all forces about any point should be equal to the moment of resultant about that point. Hence equating equations (*i*) and (*ii*), we get

$$F_1 \times AO + F_2 \times BO = F_1 \times CO - F_2 \times CO$$

or
$$F_2(BO + CO) = F_1(CO - AO)$$

$$F_2 \times BC = F_1 \times AC \qquad \begin{pmatrix} \because BO + CO = BC \text{ and} \\ CO - AO = AC \end{pmatrix}$$

or
$$\frac{BC}{AC} = \frac{F_1}{F_2} \text{ or } \frac{F_1}{F_2} = \frac{BC}{AC}$$

But $F_1 > F_2$ hence BC will be more than AC. Hence point C liles outside of AB and on the same side as the larger force F_1. *Thus in case of two unlike parallel forces the resultant lies outside the line joining the points of action of the two forces and on the same side as the larger force.*

The location of the point C, at which the resultant R is acting, can also be determined by taking moments about point A, of Figure 1.34. As the force F_1 is passing through A, the moment of F_1 about A will be zero.

The moment of F_2 about $A = F_2 \times AB$ (clockwise) (–)

Algebraic sum of moments of F_1 and F_2 about A

$$= O + F_2 \times AB = F_2 \times AB \text{(clockwise)(−)} \quad (i)$$

The moment of resultant R about A should be equal to the algebraic sum of moments of F_1 and F_2 (*i.e.*, $= F_2 \times AB$) according to the principle of moments. Also the moment of resulstant R about A should also be clockwise.

As R is acting [$\because F_1 > F_2$ and $R = (F_1 - F_2)$ so R is acting in the direction of F_1] the moment of resultant R about A would be clockwise only if the points C is towards the left of point A. Hence the point C will be outside the line AB and on the side of F_1 (*i.e.*, larger force).

Now the moment of resultant R about A

$$= R \times AC \text{(clockwise)(−)} \quad (ii)$$

Equating equations (*i*) and (*ii*),

$$F_2 \times AB = R \times AC$$

$$= (F_1 - F_2) \times AC \qquad (\because R = F_1 - F_2)$$

As F_1, F_2 and AB are known, hence AC can be calculated. Or in other words, the location of point C is known.

Resultant of two unlike parallel forces which are equal in magnitude

Two equal and opposite parallel forces acting on a body, at

some distance apart, forms a couple. Couple has the tendency to rotate the body. The perpendicualr distance between the parallel forces is called as *arm of the couple.*

Figure 1.35 shows a body on which two parallel forces which are equal in magnitued but opposite in direction are acting. These two forces will form a couple that will have a tendency to rotate the body in clockwise direction. The moment of the couple is the product of either one of the forces and perperdicular distance between the forces.

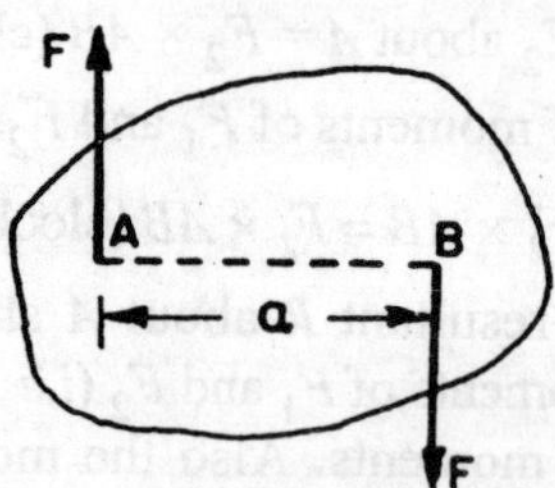

Figure 1.35

Let F = Force at A or at B

a = Perpendicular distance (or arm of the couple)

The moment (M) of the couple is given by, $M = F \times a$.

The units of moment will be Nm.

Resolution of a force into a force and a couple

A given force F applied to a rigid body at any one point say A can be replaced by an equal force at another point B together with a couple which is equal to the original force. This can be proved as :-

Let the given force F is acting at point A as shown in Figure 1.36 (*a*).

This force is to be replaced at the point B.Introduce two equal and opposite forces at B, each of magnitude F and acting parallel to the force at A as shown in Figure 1.36 (*b*). The force system of Figure 1.36 (*b)* is equivalent to the single force acting at A of Figure 1.36 (*a*). In Figure 1.36 (*b*) three equal forces are acting. The two forces *i.e.* force F at A and the oppositely directed force

F at B (*i.e.*, vertically downward force at B) from a couple. The moment of this couple is $F \times x$ clockwise where x is the perpendicular distance betweenm the lines of action of forces at A and B. The third force is acting B in the same dirction in which the forces at A is acting. In Figurc 1.36 (*c*), the couple is shown by curved arrow with symbol M. The forces system of Figure 1.36 (*c*) is equivalent to Figure 1.36 (*b*). Or in other words the Figure 1.36 (*c*) is equivalent of Figure 1.36 (*a*). So, we see that the given force F acting at A has been replaced by an equal and parallel force applied at point B in the same direction together with a couple of moment $F \times x$.

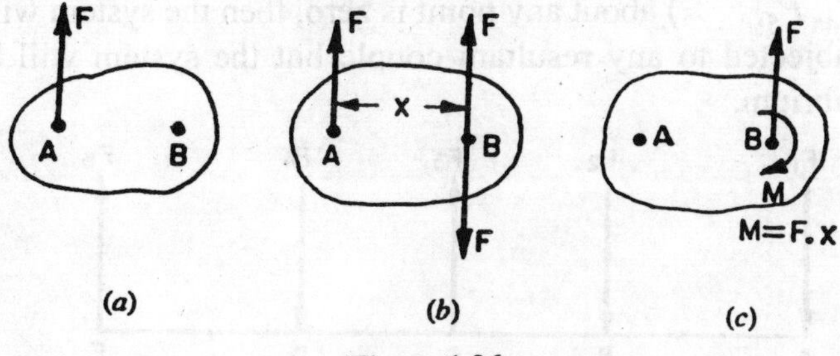

Figure 1.36

Thus a force acting at a point in a rigid body can be replaced by an equal and parallel force at any other point in the body, and a couple.

General case of parallel forces in a plane

Figure 1.37 shows a number of forces are acting on a body in one plane. All the forces are parallel to each other. The forces F_1, F_2 and F_3 are acting in one direction, whereas the forces F_3 and F_5 are acting in the opposite direction. Let R_1 = resultant of forces F_1, F_2 and F_4 and R_2 = resultant of forces F_3 and F_5. The resultant R_1 and R_2 are acting in opposite direction and are parallel to each other. Now, there is the possibility of three important cases :-

1. R_1 may not be equal to R_2. Then we shall have two unequal parallel forces acting in the opposite direction. The resultant R of these two forces (R_1 and R_2) can be easily obtained. By equating the moment of R about any point to the algebraic sum of the

moments of individual forces about the same point, the point of application of resultant R can be obtained,

2. If R_1 is equal to R_2, then we shall have two equal parallel forces (R_1 and R_2) acting in the opposite direction. The resultant R of these two forces will be zero. Now the system may reduce to a couple or the systems is in equilibrium. To distinguish between two cases, the algbraic sum of moments of all forces (F_1, F_2, ...F_5) about any point is calculated. It the sum of moments is not zero, thw system reduces a resultant couple. The calculated moment gives the moment of this couple.

3. R_1 is equal to R_2 and sum of moments of all forces (F_1, F_2, F_3, F_4, F_5, ...) about any point is zero, then the system will not be subjected to any resultant couple but the system will be in equilibrium.

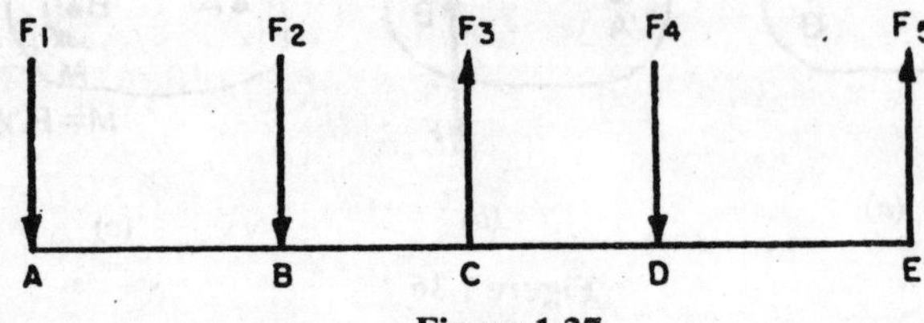

Figure 1.37

Equivalent system

In a given system of coplanar forces, the combination of a force passing through a given point and a moment about that point forms an equivalent system. The force will be the resultant of all forces which are acting on the body and the moment will be the sum of all the moments about that point.

Hence equivalent system consists of :

(*i*) a single force R passing through the given point P and

(*ii*) a single moment M_R

where R = the resultant of all force acting on the body.

M_R = sum of all moment of all the forces about P.

Funicular polygon

The resultant of a number of forces acting on a body in a plane is determined graphically by using funicular polygon method.

Let there be three forces F_1, F_2 and F_3 which are non-concurrent and coplanar acting on a body as shown in Figure 1.38. The *funicular polygon* is constructed as:

(*i*) First of all the different forces (*i.e.*,F_1, F_2 and F_3) are named by two capital letters each. The two capital letters are placed in space on either side of the force as shown in Figure 1.38 (*a*). Thus type of notation is called as *Bow's notation.*

(*ii*) In the force diagram shown in Figure 1.38 (*b*), the force vector F_1 is represented by corresponding small letters *i.e.*,the force vector F_1 is shown by *ab*. Here *ab* is parallel to F_1 *i.e.* force *AB* and equal to magnitude of F_1 to a suitable scale. Similarly *bc* is parallel of force F_2 (*i.e.*, force *BC*) and *cd* is parallel to force F_3 (*i.e.*force *CD*). Also *bc* and *cd* are equal to the magnitude of F_2 and F_3 respectively. The resultant of the force F_1, F_2 and F_3 is given by closin side *ad* in magnitude and direction. But the position of the line of action of the resultant force is unknown. To determine the point of application of the resultant force on the body, we take any arbitrary point *o*, called a pole, in the plane of the polygon of forces. (The point *o* may be taken inside or outwide of the force polygon). The points *a, b, c,* and *d* are joined to the pole *o*. The lines *ao, bo, co* and *do* ar called rays.

To determine the line of action of the resultant, we further proceed as :

(*iii*) Take any point *A* in space *A* in Figure 1.38 (*a*) (*i.e.*, to the left of line of action of action of force F_1) form *A*, draw a line *A*-1 parallel to ray *ao*. Cutting the line of action F_1 at 1. From point 1, draw a line 1-2 parallel to ray *bo* cutting the line of action F_2 at 2. From point 2, draw a line 2-3 parallel to ray *co*, cutting the line of aciton of force F_3 at 3. From point 3, draw a line 3-*D* parallel to ray *do*. In figure 1.38 (*a*), the polygon *A*-1-2-3-*D* is known as *funicular polygon* or *string polygon.*

To determine the line of action of the resultant, in Figure 1.38 (*a*), produces lines *A*-1 and *D*-3 (*i.e.* firsst line and lat line of the funicular polygon). The intersection of *A*-1 and *D*-3 determine the point 4 through which passes the line of aciton of the resultant. From point 4, draw a line parallel to ad, which is the line of action of the resultant.

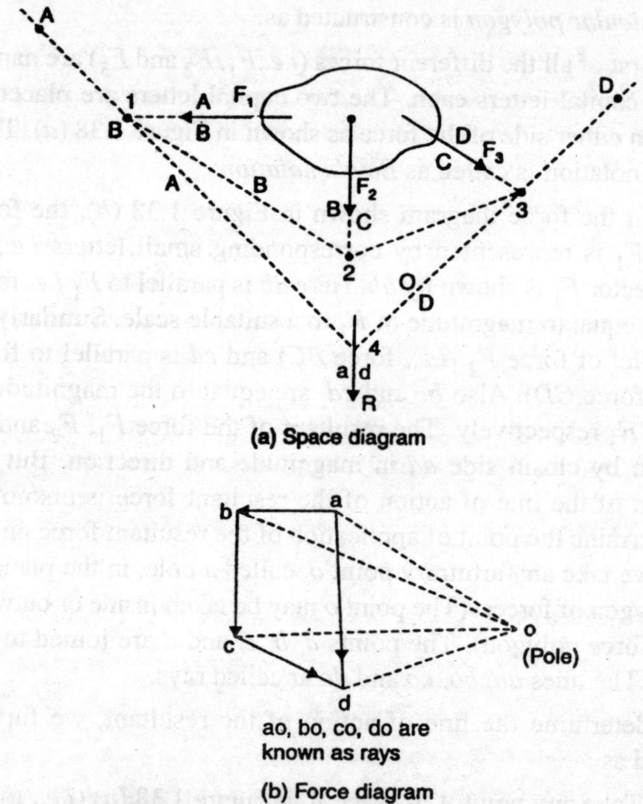

Figure 1.38 : *Funicular polygon*

Important points

1. The resultant of a force system is completely determined by constructing a force polygon and a funicular polygon. The force-polygon determines the magnitude and direction of the resultant whereas the line of action is obtained from funicular polygon of string polygon.
2. If the force polygon closes but string polygon does not close, this will be a case of a resultant couple acting on the force system.

3. The resultant force will be zero, if the force polygon closes. In that case the string polygon will also close.

CONDITIONS OF EQUILIBRIUM AND FREE BODY DIAGRAM

When we apply some external force whether concurrent or parallel, on a stationary body, the body under the influence of force may either start moving or rotating about any point. But is the body does not start moving and also does not start rotating about any point, then the body is said to be in equilibrium,. In this part the conditions of equilibrium for concurrent forces (*i.e.,* forces meeting at a point) and for parallel forces will be discussed. Also the concept of free body diagram, different types of support reactions and detetrmination or reactions will be dealt with.

Principle of equilibrium

" A stationary body when subjected to coplanar forces, will be in equilibrium if the algebraic sum of all the external forces as well as moments of all these external forces about any point in their plane is zero". This is the principle of equilibrium. Mathematically, it is expressed by the equation :

$$\Sigma F = 0 \quad (1.13)$$

$$\Sigma M = 0 \quad (1.14)$$

The sign Σ is known as sigma which is a Greek letter. This sign represents the *algebraic sum* of forces or moments.

The equation (1.13) is also known as force law of equilibrium whereas the equation (1.14) is known as moment law of equilibrium.

The forces are generally resolved into horizontal and vertical components. Hence equation (1.13) is written as

$$\Sigma F_x = 0 \quad (1.15)$$

and

$$\Sigma F_y = 0 \quad (1.16)$$

where ΣF_x = Algebraic sum of all horizontal components

ΣF_y = Algebraic sum of all vertical components.

Equations of equilibrium for non-concurrent forces systems

If the resultant of all forces and moments is zero in a non-concurrent force system, then the system will be in equilibrium.

Hence the equation of equilibrium are

$$\Sigma F_x = 0, \Sigma F_y = 0 \text{ and } \Sigma M = 0$$

Equation of equilibrium for concurrent force system: In the concurrent force system, the line of action of all forces meet at a point, and hence the moment of these force about that very point will be zero or $\Sigma M = 0$ automatically.

Thus, for concurrent force system, the condition $\Sigma M = 0$ becomes imaplicable and only two conditions i.e., $\Sigma F_x = 0$ and $\Sigma F_y = 0$ are required.

Force Law of Equilibrium

Force law of equilibriun is given by equation (1.13) or by equations (1.15) and (1.16). Now we shall apply this law to the fullowing important force system :

(*i*) Two force system.

(*ii*) Three force system.

(*iii*) Four force system.

Two force system

When a body is subjected to two forces and if these forces are collinear, equal and opposite then the body will be in equilibrium.

If the two forces acting on a body are equal and opposite but are parallel, as shown in Figure 1.39 (*a*), then the body will not be in equilibrium. This is due to the fact that the three conditions of equilibrium will not be sarisfied. This is proved as given below:

(*i*) Here $\Sigma F_x = 0$ as there is no horizontal force acting on the body. Hence Ist condition of equilibrium is satisfied.

(*ii*) Also here $\Sigma F_y = 0$ as $F_1 = F_2$. Hence second condition of equilibrium is also satisfied.

(*iii*) ΣM about any point should be zero. The resultant moment about point A is given by

$$M_A = -F_2 \times AB$$

(– ve sign is due to clockwise moment)

But M_A is not equal to zero. Hence the third condition is not satisified.

Hence a body will not be in equilibrium under the action of two equal and opposite parallel forces.

Two equal and opposite parallel forces produces a couple and moment of the couple is $-F_1 \times AB$ [see figure 1.39(a)]

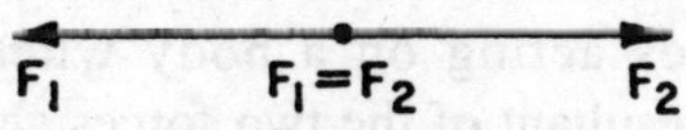

Figure : 1.39

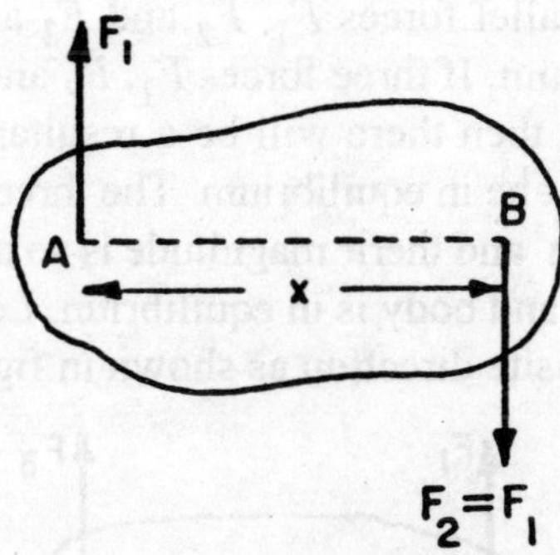

Figure : 1.39 *(a)*

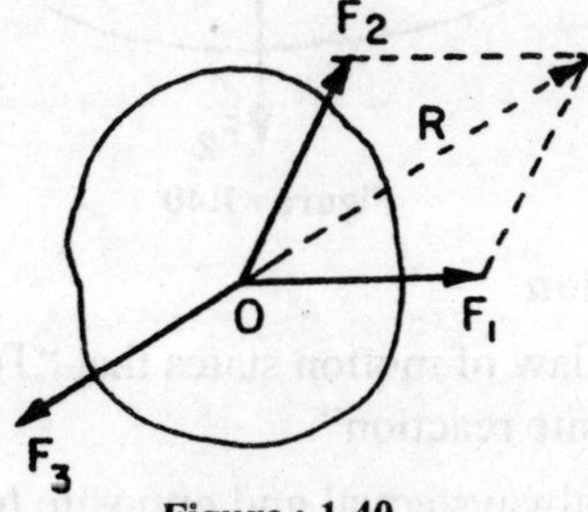

Figure : 1.40

Three force systems

When a body is subjected to three forces, which is in equilibrium then these forces may be either concurrent or parallel.. Let us first consider that the body is in equilibrium when three forces, action on the body, are concurrent. This is shown in Figure 1.40.

(*a*) *When three forces are concurrent.* The three concurrent forces F_1, F_2 and F_3 are acting on a body at point O and the body is in equilililbrium. The resultant of F_1 and F_2 is given by R. If the force F_3 is collinear, equal and opposite to the resultant R, then the body will be in equilibrium. The force F_3 which is equal and opposite to the resultant R is called as *equilibrant.* Hence for three concurrent forces acting on a body when the body is in equilibrium, the resultant of the two forces should be equal and opposite of the third force.

(*b*) *When three forces are paralle.* Figure 1.41 shows a body on which three parallel forces F_1, F_2 and F_3 are acting and the body in in equilibrium. If three forces F_1, F_2 and F_3 are acting in the same direction, then there will be a resultant $R = F_1 + F_2 + F_3$ and body will not be in equilibrium. The three forces are acting in opposite direction and therir magnitude is so adjusted that there is no resulant force and body is in equilibriun. Let us suppose that F_2 is acting in opposite direction as shown in figure 1.41.

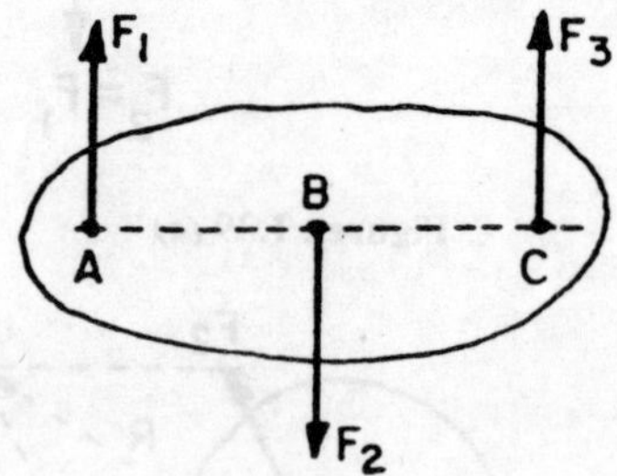

Figure : 1.40

Action and reaction

Newton's third law of motion states that "To every action there is equal and opposite reaction".

So, reaction is always equal and opposite to the action. Figure

1.42 shows that a ball which is placed on a horizontal surface (or horizontal plane) such that it is free to move along the plane but cannot move vertically downward. So the ball will exert a force vertically downwards at the support as shown in Figure 1.42 *(b)*. This force is known as action. The support will exert an equal force vertically upwards on the ball the point of contact which is known as *reaction*. Hence *any force on a support causes an equal and opposite force from the support so that action and reaction are two equal and opposite forces.*

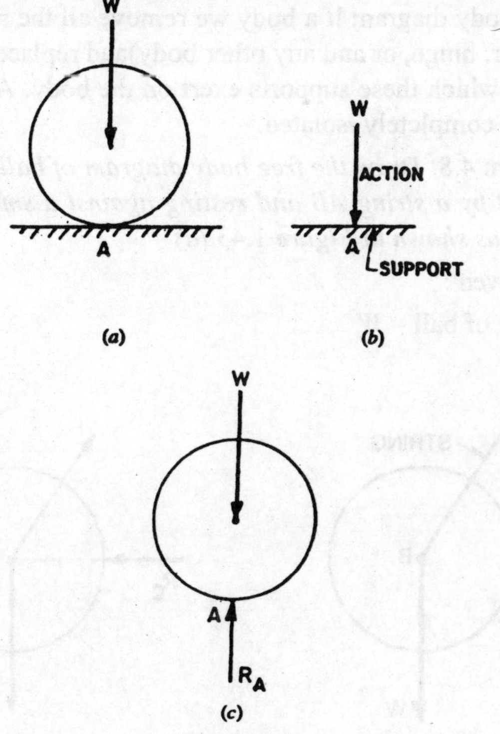

Figure 1.42

Free body diagram

The equilibrium of the bodies which are placed on the supports can be considered if we remove the supports and replace them by the reaction which ther exert on the body. In figure (1.42*a*), if we remove the supporting surface and replace it by the reaction R_A

that the surface exerts on the balls as shown in figure 1.42(c) we shall get the *free-body diagram.*

The point of application of the reaction R_A will be the point of contact A, and from the law of equilibrium of two forces, we conclude that the reaction R_A must be vertical and equal to the weight W.

Hence figure 1.42 (c), in which the ball in *completely isolated* from its support and in which all forces acting on the ball are shown by vectors, is known a free-body diagram.. Hence to draw the free-body diagram lf a body we remove all the supports (like wall, floor, hinge, or and any other body)and replace them by the reactions which these supports exert on the body. Also the body should be completely isolated.

Problem 4.8. *Draw the free body diagram of ball of weight W supported by a string AB and resting against a smooth vertical wall at C as shown in Figure* 1.43 *(a)*

Sol. Given :

Weight of ball = W

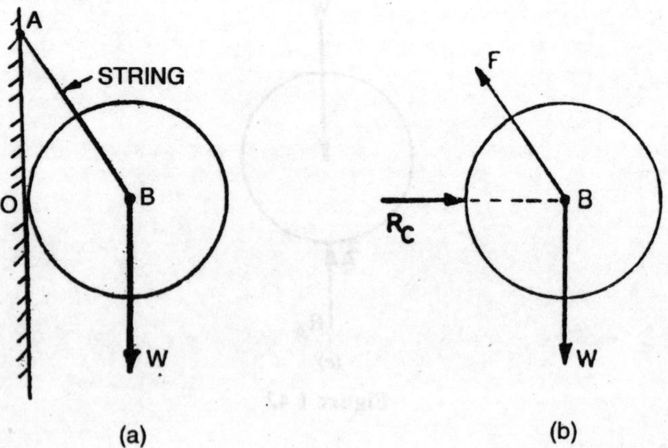

Figure 1.43

The ball is supported by a string AB and is resting aganist a vertical wall at C.

To draw the free-body diagram of the ball, isolate the ball

completely (*i.e.,* isolate the ball from the support and string).Then besides the weight *W* acitng at *B*, we have two reactive forces to apply one replacing the string *AB* and another replacing the vertical wall *AC*. Since the string is attached to the ball at *B* and since a string can pull only along its length, we have the reactive force *F* applied at *B* and parallel to *BA*. The magnitude of *F* is unknown.

The reaction R_C will be acting at the point of contact of the ball with vertical wall *i.e.* at point *C*. As the surface of wall in perfectly smooth, the reaction R_C will be normal to the vertical wall (*i.e.* reaction R_C will be horizontal in this case) and will pass through the point *B*. The magnitude of R_C is alos unknown. The complete free-body diagram is shown in Figure 1.43*b*.

2

Beams and Cables

Introduction

In this chapter we shall discuss about the analyis of the internal forces in two important types of engineering structures, namely,

1. *Beams*, are usually long, straight prismatic members which are designed to support loads applied at various points along the memeber.
2. *Cable*, are flexible members capable of withstanding only tension and are designed to support either concentrated or distributed loads. Cables are used in many engineering applicaitions, such as suspension bridges and transmisionlines

Internal forces in members

Let us first consider a straight *two-force member AB* (figure 2.1a).Forces F and -F are acting at *A* and *B*, respectively also directed along AB in opposite sense and have the same magnitude. Now, let us cut the member at *C*. To maintain the equilibrium of the free bodies *AC* and *CB* thus obtainded, we must apply to *AC* a force –**F** equal and opposte of **F**, and to *CB* a force **F** equal and opposite of –**F** (Figure 2.1b). These new forces are directed along *AB* in opposte sence and have the same magnitude *F*. Before the members was cut, the two parts *AC* and *CB* were in equilibrium wihich mean an internal force equivalent to these new forces must have existed in the member itself. We conclude that in the case of a straight two-force member, the internal forces that the two portions of the member exert on each other are equivalent to *axial forces*. The commom magnitude *F* of these forces does not depend upon the the location of the section *C* and is '1' reffered to as the *force in member AB*. In the case considered, the member is in tension and will elongate under the action of the internal forces. In the case represented in Figure 2.2, the member is in compression and will decrease in length under the action of the internal forces

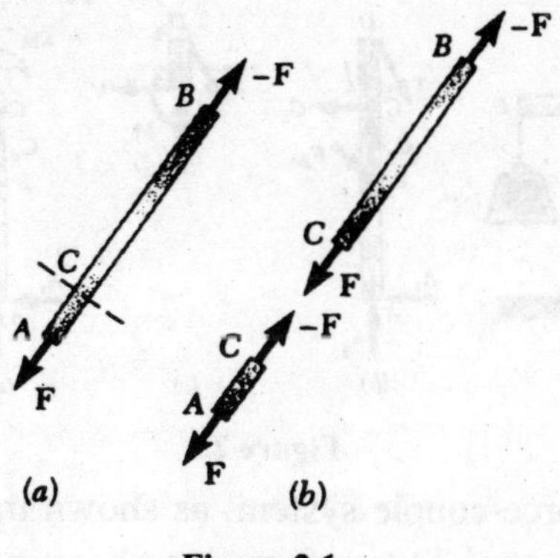

Figure 2.1

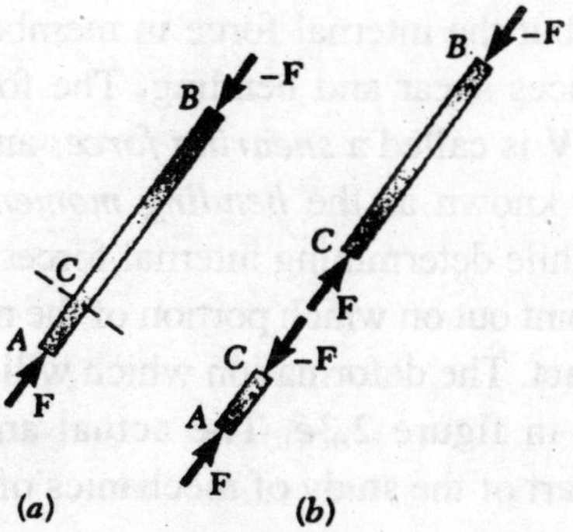

Figure 2.2

Now, let us consider a *multiforce member.* For example, A crane is shown in Figure 2.3a and the free body diagram of member *AD* is drawn in figure 2.3b. We now cut member *AD* at *j* and draw a free-body diagrame for each of the portions *JD* and *AJ* of the member (Figure 2.3c and d). To maintain the equilibrium of free body *JD* we have to apply at *J* a force *F* to balance the vertical component of *T*, a force *V* to balance the horizontal componet of **T**, and a couple **M** to balance the moment of **T** about *J*. Again we conclude that internal forces must have existed at *J* before member *AD* was cut. The internal forces acting on the portion *JD* of member *AD* are equivalent ot the force-couple system shown in figure 2.3c. According to Newton's third law, the internal forces acting of *AJ* must be equivalent to an equal

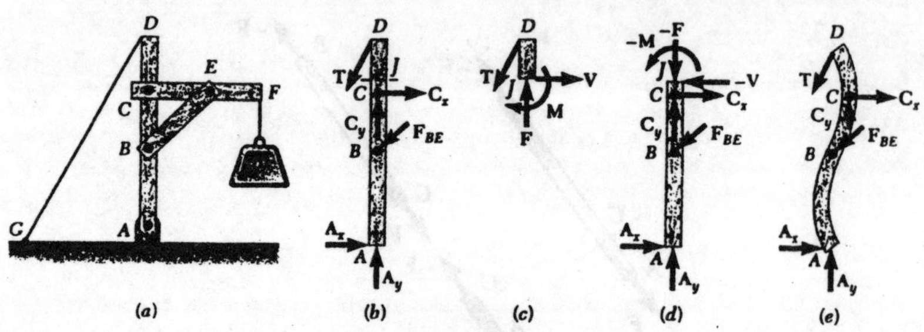

Figure 2.3

and opposite force-couple system, as shown in figure 2.3d. The internal force in straight two force member produce only tension or compression but the internal force in member *AD* along with these also produces shear and bending. The force **F** is an *axial force;* the force **V** is called a *shearing force;* and the moment **M** of the couple is known as the *bending moment at J.* It is very important that while determining internal forces in a member, we should clearly point out on which portion of the member the forces are supposed to act. The deformation which will occur in member *AD* is sketched in figure 2.3*e*. The actual analysis of such a deformation is part of the study of mechanics of materials.

The internal forces are also equivalent to a force-couple system in a two-force member which is not straight. This is shown in figure 2.4, where the two-force member *ABC* has been cut at *D*.

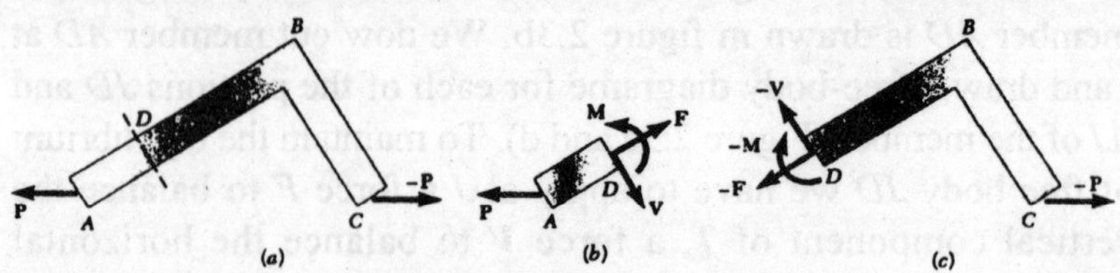

Figure 2.4

BEAMS

Various types of loading and support

A structural member which is designed to support loads applied at various points along thc member is known as *beam*. In most cases, the loads are perpendicular to the axis of the beam and will cause only shear and bending in the beam. Loads also produced axial forces in the beam, when they are not at a right angle to the beam.

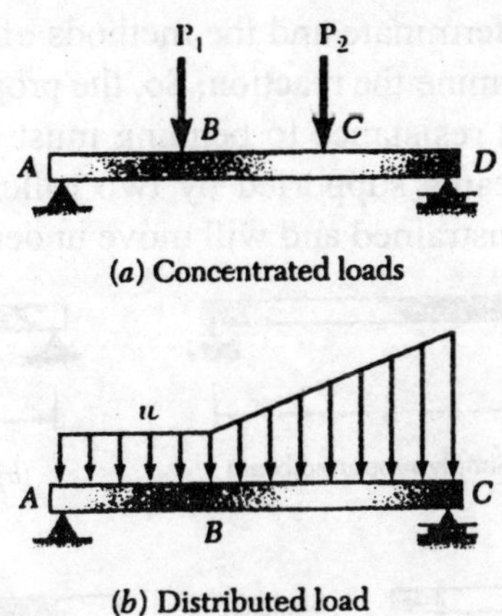

(*a*) Concentrated loads

(*b*) Distributed load

Figure 2.5

Beams are usually long, straight prismatic bars. Designing *a* beam for the most effective support of the applied loads is a two-part precess: (1) determining the shearing forces and bending moments produced by the loads and (2) selecting the cross section best suited to resist the shearing forces and bending moment determined in the first part. Here we are concerned with the first part of the problem of beam design. The second part belongs to the study of mechanics of materials.

A beam can be subjected to *concentrated loads* $\mathbf{P}_1$, $\mathbf{P}_2$, . . . , pressed in newtons or kilonewtons (figure 2.5*a*), or to a combination of both. When the load *w* per until length has a constant value over part of the beam (jas between *A* and *B* in figure 2.5*b*), the load is said to be *uniformly distributed* over that part of the beam. The determination of the reaction at the supports in considerably simplified if distributed loads are replaced by equivalent concentrated loads. This substitution, however should not be performed, or at least should be performed with care, when internal forces are being computed.

Beams are classified according to the way in which they are supported. Several types of beams frequently used are shown in Figure 2.6. The distance L between supports is called the *span*. If the support involve only three unknown the reaction will be determinate while if more unknown are involved, the reaction will be statically indeterminate and the methods of statics will not be sufficient of determine the reaction. So, the properties of the beam with regard to its resistance to bending must then be taken into consideration. Beams supported by two rollers are (not shown here) partially constrained and will move under certain loadings.

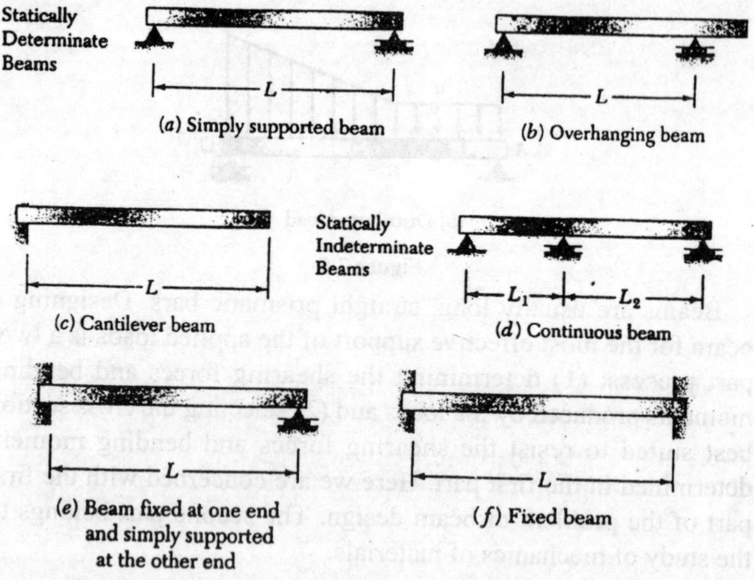

Figure 2.6

Sometimes two or more beams are connected by hinges to form a single continuous structure. Two examples of beams hinged at a point H are shown in figure 2.7. It will be noted that the reactions at the supports involve four unknown and cannot be determined from the free-body diagram of the two-beam system. They can be determined, however, by considering the free-body diagram of each beam separately, six unknown are involved (including two force components at the hinge), and six equations are available.

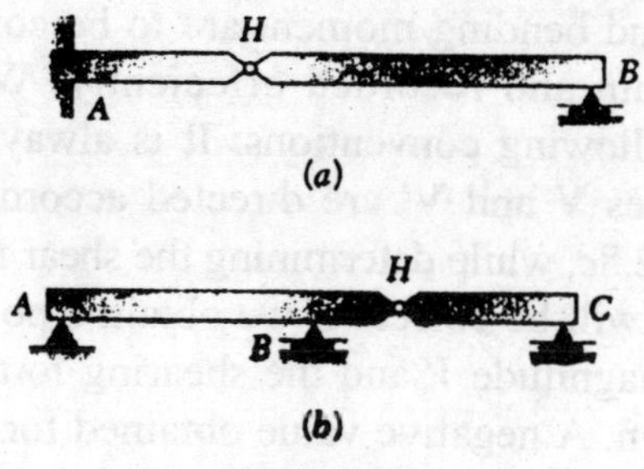

Figure 2.7

Shear and bending moment in a beam

Consider a beam *AB* which is subjected to various concentrated and distributed loads (figure 2.8*a*). We are supposed ot determine the shearing force and bending moment at any point of the beam. In this example, the beam is simply supported, but the method used could be applied to any type of statically determinate beam.

First we determine the reactions at *A* and *B* by choosing the entire beam as a free body (figure 2.8*b);* writing $\Sigma M_A = 0$ and $\Sigma M_B = 0$, we obtain, respectively, $\mathbf{R}_B$ and $\mathbf{R}_A$.

To determine the i nternal forces at *C*, we cut the beam at *C* and draw the free-body diagrams of the portions *AC* and *CB* of the beam (figure 2.8*c*) with the help of free body diagram of *AC*, we can determine the shearing force **V** at *C* by equating to zero the sum of the vertical components of all forces acting on *AC*. Simi!arly, the bending moment **M** at *C* can be found by equating to zero the sum of the moments about *C* of all forces and couples acting on *AC*. Alternatively, we could use the free-body diagram of *CB* and determine the shearing force **V′** and the bending moment **M′** by equating to zero the sum of the vertical components and the sum of the moments about *C* of all forces and couples acting of *CB*. While this choic of free bodies may facilitate the computation of the numerical values of the shearing force and bending moment, it makes it necessary to indicate on which portion of the beam the internal forces considered are acting. It is essential to find a way to avoid having to specify every tme which portion of the beam is used as a free body, if the shearing force and bending moment are to be computed at very point of the beam and recorded effeciently. We shall adopt, therefore, the following conventions: It is always assumed that the internal forces **V** and **V′** are directed according to the way shown in figure 2.8*c*, while determining the shear force in a beam. Our assumption will be correct if we obtain a positive value for their common magnitude *V* and the shearing forces are actually directed as shown. A negative value obtained for *V* will indicate that the assuption was wrong and that the shearing forces are directed in the opposite way. Thus, only the magnitude *V*, together with a plus or minus sign, nedds to be recorded to define completely the shearing forces at a given point of the beam. The shear at the given point of the beam is commonly reffered by the scalar *V*.

In the same way it *will always be assumed* that the internal comples **M** and **M′** are directed as shown in figure 2.8*c*. If bending momemt commonly reffered by *M* has the postive value obtained for their magnitude indicates that this assuption was correct, and a negative value will indicate that it was wrong. Using, the sign conventions we have presented, we summerize:

The shear V and the bending moment M at a given point of a beam are said to be positive when the internal forces and couples acting on each portion of the beam are directed as shown in figure 2.9a.

These conventions can be more easily remembered if we note that:

1. *When the **external** forces (loads and reactions) acting on the beam tend to shear off the beam at C as indicated in Figure 2.9b, the shear at C is positive.*
2. *When the **external** forces acting on the beam tend to bend the beam at C as indicated in Figure 2.9c, the bending moment at C is positive.*

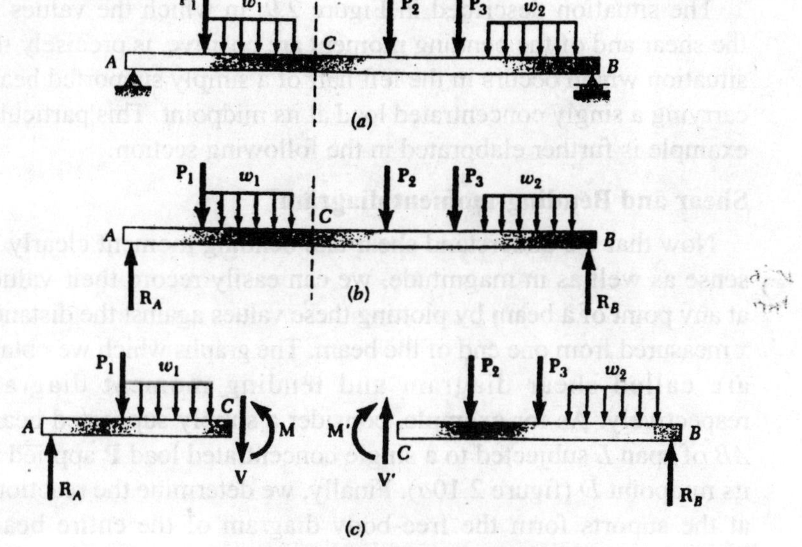

Figure 2.8

(*a*) Internal forces at section
(positive shear and positive bending moment)

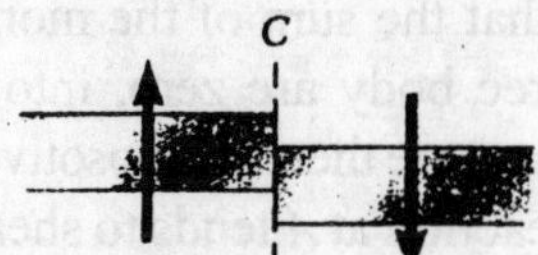

(*b*) Effect of external forces
(positive shear)

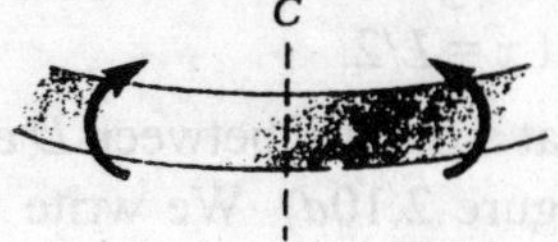

(*c*) Effect of external forces
(positive bending moment)

Figure 2.9

The situation described in Figure 2.9, in which the values of the shear and of the bending moment are positive, is precisely the situation which occurs in the left half of a simply supported beam carrying a singly concentrated load at its midpoint. This particular example is further elaborated in the following section.

Shear and Bending-moment diagram

Now that we understand shear and bending moment clearly in sense as well as in magnitude, we can easily record their values at any point of a beam by plotting these values against the distance x measured from one end of the beam. The graphs which we obtain are called shear diagram and tending moment diagram respectively. As for example, consider a simply supported beam AB of span L subjected to a single concentrated load **P** applied at its midpoint D (figure 2.10*a*). Finally, we determine the reactions at the suports form the free-body diagram of the entire beam (figure 2.10*b*) and we find that the magnitude of each reaction is equal to $P/2$.

Now, next we cut the beam at a point C between A and D and draw the free-body diagrams of AC and CB (figure 2.10*c*). We direct the internal forces **V** and **V′** and the internal couples **M** and **M′** as indicated in figure 2.9*a* assuming that shear and bending moment are postive. We find $\mathbf{V} = +P/2$ and $\mathbf{M} = +Px/2$ by taking the free body AC and writing that the sum of the moments about C of the forces acting on the free body are zero, into consideration. Both shear and bending moment are therefore posotive; this can be checked by observing that the reaction at A tends to shear off and to bend the beam at C as indicated in figure 2.9*b* and *c*. We can plot V and M between A and D (figure 2.10*e* and *f*); the shear has a constant value $V = -P/2$, while the bending moment increases linearly from $M = 0$ at $x = 0$ to $M = PL/4$ at $x = L/2$.

Now, cut the beam at a point E between D abnd B and consider the free body EB. (Figure 2.10*d*). We write that the sum of the vertical components and the sum of the moments about E of the forces acting on the free body are zero. we obtain $V = -P/2$ and $M = P\ (L - x)/2$. We obtain a negative shear and positive bending moment, this is checked by observing that the reaction at B bends

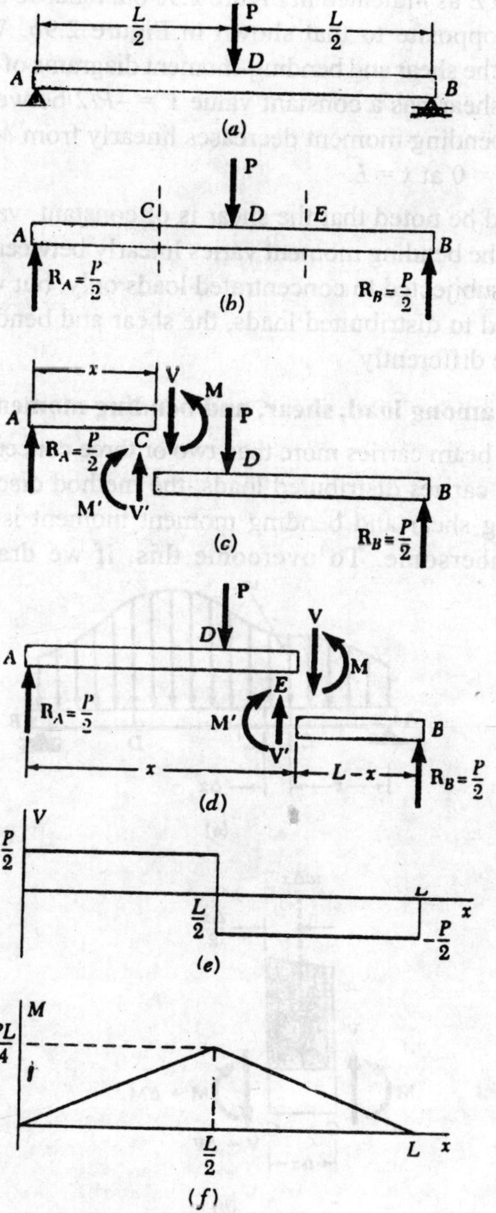

L/2
P
L/2
D
A
B
(a)
P
C
D
E
A
B
R_A = P/2
(b)
R_B = P/2
x
V
M
A
P
R_A = P/2
C
D
M'
V'
B
(c)
R_B = P/2
P
V
D
A
M
E
R_A = P/2
M'
V'
B
x
L − x
R_B = P/2
(d)
V
P/2
L
x
L/2
−P/2
(e)
M
PL/4
L/2
L
x
(f)

Figure 2.10

the beam at E as indicated in Figure 2.9*c* but tends to shear it offon a manner opposite to that shown in Figure 2.9*b*. We can now complete, the shear and bending-moment diagrams of Figure 2.10*e* and *f*; the shear has a constant value $V = -P/2$ between D and B, while the bending moment decreases linearly from $M = PL/4$ at $x = L/2$ to $M = 0$ at $x = L$.

It should be noted that the shear is of constant value between loads and the bending moment varies linearly between loads when a beam is subjjected to concentrated loads only, but when a beam is subjected to distribuited loads, the shear and bending moment vary quirte differently.

Relations among load, shear, and bending moment

When a beam carries more than two or three concentrated loads, or when it carries distributed loads, the method discribed earlier for plotting shear and bending moment moment is likely to be quite cumbersome. To overcome this, if we draw a certain

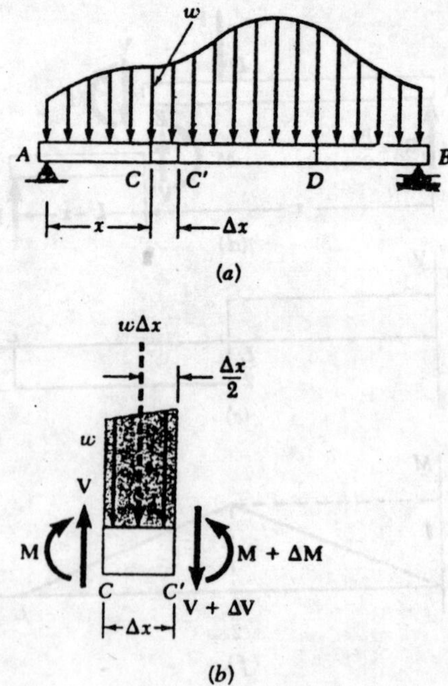

Figure 2.11.

relationship between load, shear and bending moment than the construction of shear diagram and especially of the bending moment diagram will be greatly facilitated.

Consider a simply supported beam *AB* carrying a distributed load *w* per unt length (figure 2.11*a*), and let *C* and *C′* be two points of the beam at a distance Δx from each other. The shear and bending moment at *C* are assumed to be positive and are denoted by *V* and *M* respectively. The shear and bending moment at *C′* are dented by $V + \Delta V$ and $M + \Delta M$.

Now, let us detach the portion of beam *CC′* and draw its free-body diagram (figure 2.11*b*).The forces exerted on the free body include a load of magnitude $w\,\Delta x$ and internal forces and couples at *C* and *C′*. The forces and couples will be directed as shown if the figure, as shear and bending moments are assumed to be positive

Relations between load and shear

The sum of the vertical components of the forces acting on the free body *CC′* is zero:

$$V - (V + \Delta V) - w\,\Delta x = 0$$

$$\Delta V = -w\,\Delta x$$

Dividing both members of the equation by Δx and then letting Δx approach zero, we obtain:

$$\frac{dV}{dx} = -w \tag{2.1}$$

Formula (2.1) indicates that for a beam loaded as shown in figure 2.11*a*, the slope of *dV/dx* lof the shear curve is negative; the numerical value of the slope at any point is equal to the load per unit length at that point.

On, integrating (2.1) between points *C* and *D*, we obtain

$$V_D - V_C = -\int_{x_C}^{x_D} w\,dx \tag{2.2}$$

$V_D - V_C = -$ (area under load curve between *C* and *D*) (2.2′)

By considering the equilibrium of the portion of beam *CD*, same result will be obtained because the area under the load curve represents the total load applied between *C* and *D*.

At point where a concentrated load is applied the formula 2.1 is not valid, the shear curve is discountinuous at such a point. Similarly, formulas (2.2) and (2.2') cease to be valid when concentrated loads are applied between *C* and *D*, as they do not take into account the sudden change in shear caused by a concentrated load. Therefore formulas 2.2 and 2.2' should be applied only between successive concentrated loadds.

Relations between Shear and Bending Moment

Seeing the free-body diagram of Figure 2.11 (*b*), and writing the sum of the moments about *C* is zero, we obtain

$$(M+\Delta M)-M-V\Delta x+w\,\Delta x\frac{\Delta x}{2}=0$$

$$\Delta M=V\Delta x-\frac{1}{2}w(\Delta x)^2$$

Dividing both members of the equation by Δx and then letting Δx approach zero, we obtain

$$\frac{dM}{dx}=V \tag{2.3}$$

The slope *dM*/*dx* of the bending-moment curve is equal to the value of the shear as indicated by the formula 2.3. This is true at any point where the shear has a well-defined value, i.e., at any point where no concentrated load is applied. Formula (2.3) also shows that the shear is zero at points where the bending moment is maximum. This property helps in determining the points where the beams is likely to fail under bending.

Integrating (2.3) between points *C* and *D*, we obtain

$$M_D-M_C=\int_{x\,C}^{x\,D}V\;dx \tag{2.4}$$

M_D-M_C = area under shear curve between *C* and *D* (2.4')

The area under the shear curves should be considered positive where the shear is positive and negative where the shear is negative. As long as the shear curve has been correctly drawn, the formulae (2.4) and (2.4') are considered valid even when concenterated loads areapplied between *C* and *D*. The formulas cease to be valid, however, if a *couple* is applied at a point between

C and *D*, since they do not take into account the sudden change in bending moment caused by a couple.

In most engineering applications, the value of the bending moment needs to be known only at a few specific points. Once the shear diagram has been drawn, and after *M* has been determined at one of the ends of the beam. Then using formula (2.4) and computing the area under the shear curve we can obtain the value of the bending moment at any given point. For instance, since $M_A = 0$ for the beam of figure 2.12, by simply measuring the area of the shaded triangle in the shear diagram, the maximum value of the bending moment for that beam can be obtained.

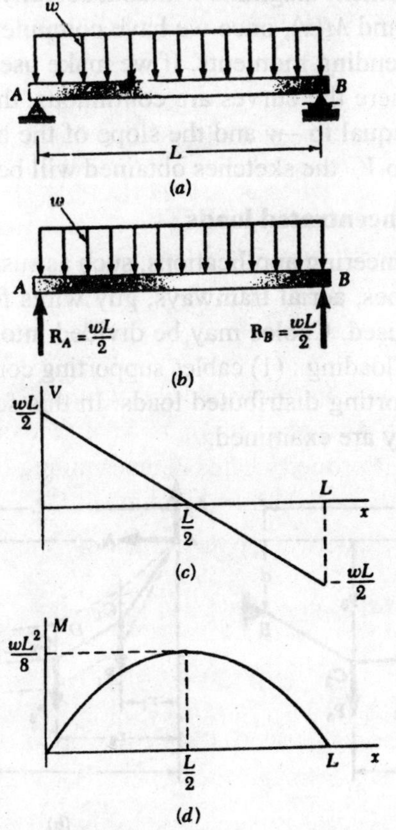

Figure 2.12

$$M_{\max} = \frac{1}{2}\frac{L}{2}\frac{wL}{2} = \frac{wL^2}{8}$$

The load curve is a horizontal straight line, the shear curve is an oblique straight line, and the bending-moment curve is a parabola, in this example. If the load curve had been an oblique straight line (first degree), the shear curve would have been a parabole (second degree), and the bending-moment curve would have been a cubic (third degree). It means the shear and bending-moment cuves will always be, respectively, one and two degress higher than the load curve. We should be able to sketch the shear and bending-moment diagrams withuout actually determining the functions $V(x)$ and $M(x)$, once we have computed a few value of the shear and bending moments. If we make use of the fact that at any point where the curves are continuous, the slope of then shear curve is equal to $-w$ and the slope of the bending-moment curve is equal to V, the sketches obtained will be more accurate.

Cables with concentrated loads

In many engineering appllications, such as suspension bridges, teansmission lines, aerial tramways, guy wires for high towwer, etc. cables are used. Cables may be divided into two catagories, according their loading : (1) cables supporting concentrated loads, (2) cables supporting distributed loads. In this section, cables of the first category are examined.

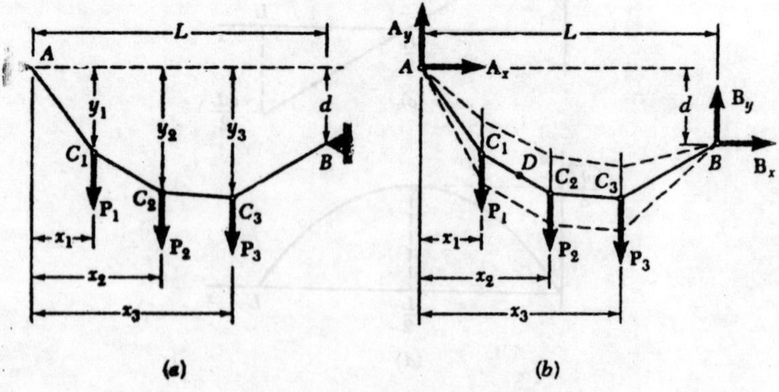

Figure 2.13

Let us consider a cable attached to two fixed points A and B and supporting n vertical concentrated loads $\mathbf{P}_1, \mathbf{P}_2, \ldots, \mathbf{P}_n$(figure 2.13$a$). We assume that its resistance to bending is small and can be neglected is the cable is flexible. We further assume that the *weight of the cable is negligible* compared with the loads supported by the cable. So, any portion of cable between succerssive loads can therefore be considered as a two-force member, and the internal forces at any point in the cable can be reduce to a *force of tension dirceted along the cable.*

Assuming that each of the loads lies in a given vertical line *i.e.* that the horizontal distance from support A to each of the loads is known; we also assume that the horizontal and vertical distances between the supports are known. We are about to determine the shape of the cable *i.e.*, the vertical distance form support A to each of the points $C_1, C_2, \ldots, C_n$, and also the tension T in each portion of the cable.

Firstly, we draw the free-body diagram of the entire cable (figure 2.13 b). The reactions at A and B must be represented by two components each as we do not know the slope of the portion of cable attached at A and B. The three equations of equilibrium are not sufficient of determine the reactions of A and B as four unknown are involved. We must therefore obtain an additional equation by considering the equililbrium of a portion of the cable.

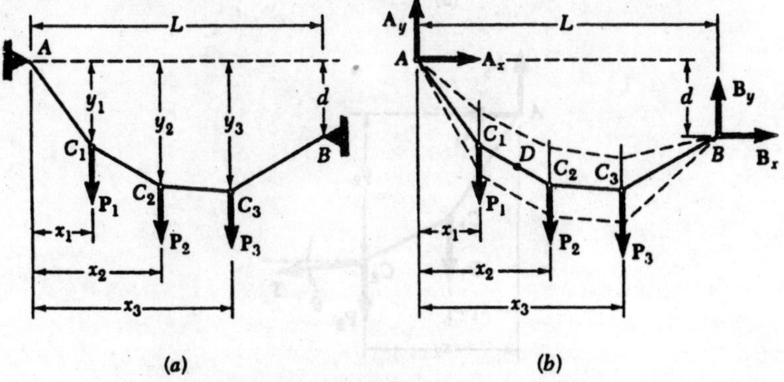

Figure 2.13 *(repeated)*

By Knowing the coordinates x and y of a point D of the cable we can make it possible. Drawing the free-body diagram of the portion of cable *AD* (figure 2.14*a*) and writing $\Sigma M_D = 0$, we get on additional relation between the scalar components A_x and A_y and can determine the reactions *A* and *B*. If we did not know the co-ordinate of *D* then our problem would remain indeterminate unless some other between A_x and A_y (or between B_x and B_y) were given. The cable might hang in various possible ways, as indicated by the dashed lines in Figure 2.13*b*.

The vertical distance from *A* to any point of the cable can easily be found once A_x and A_y have been determined. Considering point C_2, for example, we draw the free-body diagram of the portion of cable AC_2 (Figure 2.14*b*). Writing $\Sigma M_{C2} = 0$, we obtain an equation which can be solved for y_2. We observe that $T \cos \theta = -A_x$; *the horizontal component of the tension force is the same at*

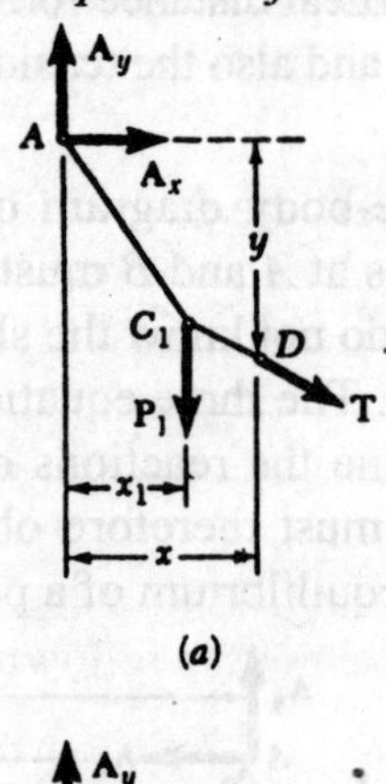

(*a*)

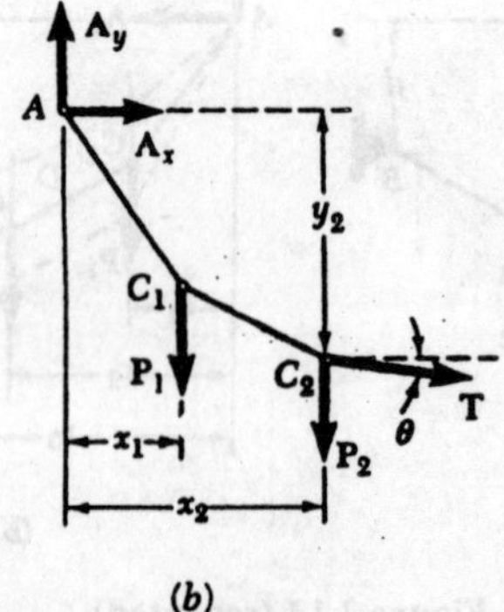

(*b*)

Figure 2.14

any point of the cable. It indicate that the tension T is maximum when $\cos\theta$ is minimum, *i.e.,* in the portion of cable which has the larges angle of inclination θ. Clearly, this portion of cable must be adjacent to one ot the two supports of the cable.

Cables with distributed loads

Now, let us consider a cable attache to two fixed point A adn B and carrying a *distributed load* (figure 2.15*a*). We saw in the preceding section that in case of a cable supporting concentrated loads, the internal force at any point is a force of tension directed along the cable but a cable which carries a distributed load, hangs in the shape of a cureve, and the internal force at a point D is a force of tension **T** *directed along the tangent to the curve.* In this section our main purpose is to learn how to determine the tension at any point of a cable supporting a given distributed load. In the following sections, the shape of the cable will be determined for two partilcular types of distributed loads.

We consider the most general case of distributed load and draw the free-body diagram of the portion of cable extending from the lowwest point C to a given point D of the cable (figure 2.15*b*). The forces acting on the free body are the horizontal tension force $\mathbf{T}_0$ at C, the tension force **T** at D, directed along the tangent to the cable at D, and the resultant **W** of the distributed load supported by the portion of cable CD. Drawing the corresponding force triangle (figure 2.15*c*), we get the following relations:

$$T\cos\theta = T_0 \qquad T\sin\theta = W \tag{2.5}$$

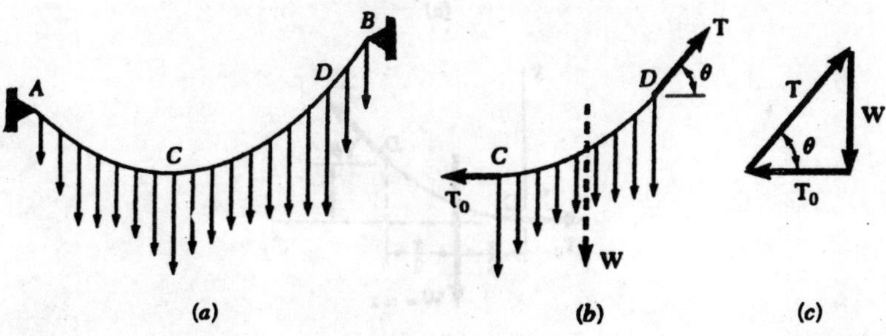

Figure 2.15

$$T = \sqrt{T_0^2 + W^2} \qquad \tan\theta = \frac{W}{T_0} \tag{2.6}$$

from the relations (2.5), it seems that the horizontal component of the tension force **T** is the same at any point and that the vertical component of **T** is equal to the magnitude W of the load measured form the lowes point. Relations (2.6) show that the tension T is minimum at the lowest point and maximum at one of the two points of support.

Parabolic cable

Now, let us assume that a load *uniformly distributed along the horozontal* (figure 2.16*a*) is carried by cable AB. Cables of suspension bridges may be supposed to be loaded in this way, since the weight of the cables is small compared with the weight of the roadway. We denote the load per unit length (*measured horizontaly*) by and express it in *N/m*. We find that the magnitude W of the total load carried by the portion of cable extending from C to the point D of co-ordinate x and y is $W = wx$ by choosing co-ordinate axes with origin at the lowest point C of the cable.

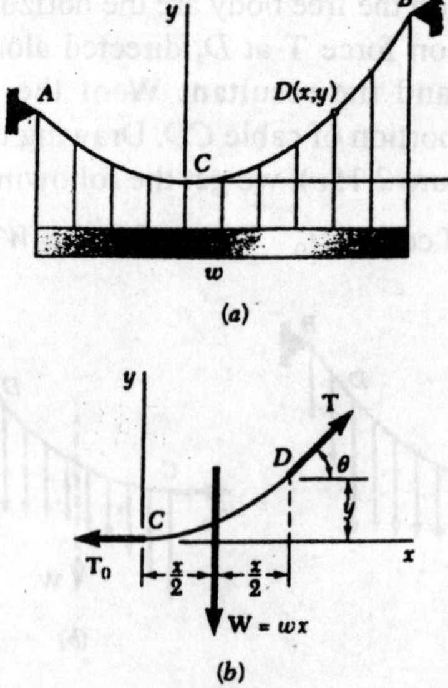

Figure 2.16

The relations (2.6) defining the magnitude and direction of the tension force at D become

$$T = \sqrt{T_0^2 + w^2x^2} \qquad \tan\theta = \frac{wx}{T_0} \tag{2.7}$$

Moreover, half the horizontal distance from C to D is equal to the distance from D to the line of action of the resultant W (figure 2.16*b*).Summing moments about D, we write

$$+\Sigma M_D = 0: \qquad wx\frac{x}{2} - T_0 y = 0 \tag{2.7}$$

and, solving for y,

$$y = \frac{wx^2}{2T_0} \tag{2.8}$$

Finally, this is the equation of a *parabola* with a verticle axis and its vertex at the origin of coordinates. The curve formed by cables loaded uniformly along the the horizontal is thus a parabole.

The distance L betwen the supports is called span of the cable when the supports A and B of the cable have the same elevation while sag of the cable is the vertical distance h from the supports to the lowest point.(figure 2.17*a*). If the span and sag of a cable are known, and if the load w per unit horizontal length is given, the minimum tension T_0 may by found by substituting $x = L/2$ and $y = h$ in equation(2.8). With the help of equation 2.7 the tension and the slope at any point of the cable can be determined and equation 2.8 gives the shape of the cable.

But, when the supports have different elevations, the positions of the lowest point of the cable is not known and the coordinates x_A, y_A and x_B, y_B of the supports must be determined. For this purpose we write that the co-ordinates of A and B satisfy equation (2.8) and that $x_B - y_A = L$ and $y_B - y_A = d$, where L and d denote, respectively, the horizontal and vertical distances between the two supports (figure 2.17*b* and *c*).

The length of the cable form its lowest point C to its support B can be calculated from the formula

$$s_B = \int_0^{x_D} \sqrt{1 + \left(\frac{dy}{dx}\right)^2}\, dx \tag{2.9}$$

On defferentation of (2.8) we obtain the derivates $dy/dx = wx/T_d$ substituting it into (2.9) and using the binomial theorem to expand the radical in an infinite series, we have

$$s_B = \int_0^{x_B} \sqrt{1 + \frac{w^2 x^2}{T_0^2}} dx$$

$$= \int_0^{x_B} \left(1 + \frac{w^2 x^2}{2T_0^2} - \frac{w^4 x^4}{8T_0^4} + \cdots\right) dx$$

$$s_B = x_B\left(1 + \frac{w^2 x^2{}_B}{6T_0^2} - \frac{w^3 x^4{}_B}{40T_0^4} + \cdots\right)$$

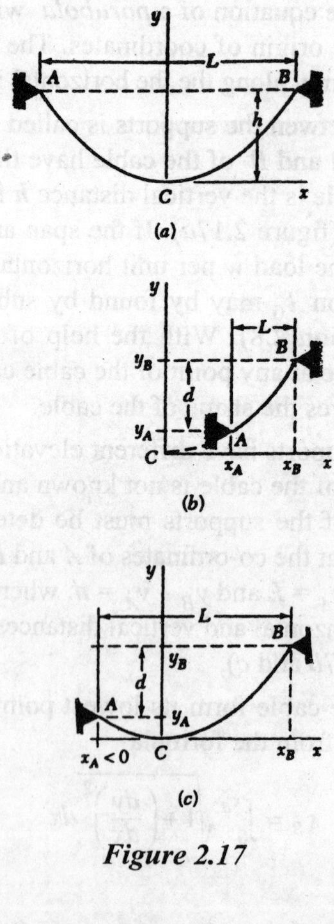

(a)

(b)

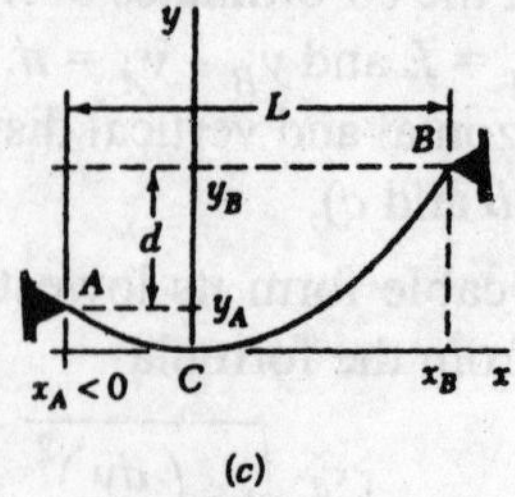

(c)

Figure 2.17

and, since $wx_B^2 / 2T_0 = y_B$,

$$s_B = x_B\left[1+\frac{2}{3}\left(\frac{y_B}{x_B}\right)^2-\frac{2}{5}\left(\frac{y_B}{x_B}\right)^4+\cdots\right] \tag{2.10}$$

The series converges for values of the valur y_B / y_B less than 0.5, only the first two terms of the series need be computed as in most cases, this ratio is much smaller.

Catenary

Let us now consider a cable *AB* carrying a load *uniformly distributed along the cable itself (*figure 2.18*a)*. This type of loading is suitable for the cables hanging under their own weight. The load unit length (measued along the cable)is denoted by *w* and express it in *N/m*.. The magnitude *W* of the total load carried by a portion of cable of length s extending from the lowest point *C* to a point *D* is $W = ws$. Substituting this value for *W* in formula (2.6), we obtain the tension at *D*:

$$T = \sqrt{T_0^2 + w^2s^2}$$

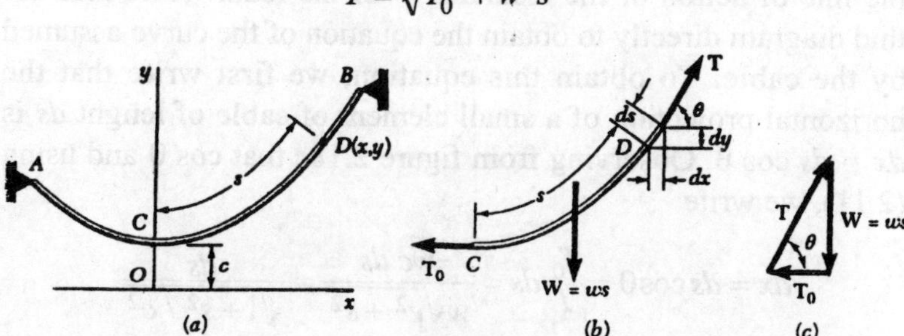

Figure 2.18.

In order to simplify the further calculation, we introduce the constant $c = T_0/w$. We thus write

$$T_0 = wc \quad W = ws \quad T = w\sqrt{c^2 + s^2} \tag{2.11}$$

This integral can be found in all standard integral tables. The function

$$z = \sinh^{-1}u$$

(read "are hyperbolic sine *u*") is the *inverse* of the function $u =$

sinh z (read "hyperbolic sine z"). This function and the function $v = \cosh z$ (read "hyperbolic cosine z") are defined as follows:

$$u = \sin\text{h}\, z = \frac{1}{2}(e^{z} - e^{-z}) \qquad v = \cosh z = \frac{1}{2}(e^{z} + e^{-z})$$

Numerical values of the function sinh z and cosh z are found in *tables of hyperbolic functions.* They may also be computed on most calculators either directily or from the above definitions. The student is refferred to any calculus text for complete descriptin of the properties of these functions. In this section, we use only the following properties, which are easily derived from the above defintions:

$$\frac{d\sinh z}{dz} = \cosh z \qquad \frac{d\cosh z}{dz} = \sinh z \tag{2.12}$$

$$\sinh 0 = 0 \quad \cosh 0 = 1 \tag{2.13}$$

$$\cosh^2 z - \sinh^2 z = 1 \tag{2.14}$$

Figure 2.18*b* shows the free body diagram of the portion of cable *CD*. As we do not know the horizontal distance from *D* to the line of action of the resultant *W* of the load, we cannot use thid diagram directly to obtain the equation of the curve assumed by the cable. To obtain this equation, we first write that the horizontal projection of a small element of cable of lenght *ds* is $dx = ds \cos\theta$. Observing from figure 2.18*c* that cos θ and using (2.11), we write

$$dx = ds\cos\theta = \frac{T_0}{T}ds = \frac{wc\, ds}{w\sqrt{c^2 + s^2}} = \frac{ds}{\sqrt{1 + s^2 / c^2}}$$

By selecting the origin *O* of the coordinates at a distance *c* directly below *C* (Figure 2.18*a*) and intergrating from *C(O, c)* to *D* (*x, y*), we get

$$x = \int_0^s \frac{ds}{\sqrt{1 + s^2 / c^2}} = c\left[\sin^{-1}\frac{s}{c}\right]_0^s = c\sinh^{-1}\frac{s}{c}$$

This equation, which relates the length *s* of the portion of cable *CD* and the horizontal distance *x*, can be written in the form

$$s = c\sinh\frac{x}{c} \tag{2.15}$$

By writing $dy = dx \tan\theta$ we can now obtain the relation between the coordinates x and y. Observing form Figure 2.18*c* that $\tan\theta = W/T_0$ and using (2.11) and (2.15), we write

$$dy = dx\tan\theta = \frac{W}{T_0}dx = \frac{s}{c}dx = \sinh\frac{x}{c}dx$$

On, integrating from $C\,(O, c)$ to $D(x, y)$ and using (2.12) and (2.13), we get

$$y - c = \int_0^x \sinh\frac{x}{c}\,dx$$

$$= c\left[\cosh\frac{x}{c}\right]_0^x = c\left(\cosh\frac{x}{c} - 1\right)$$

$$y - c = c\cosh\frac{x}{c} - c$$

which we reduces to

$$y = c\cosh\frac{x}{c} \tag{2.16}$$

This is the equation of a *catenary* with vertical axis The parameter of a catenary is the ordinate c of the lowest point C. Squaring both sides of Equations (2.15) and (2.16), subtracting, and taking (2.14) into account, we obtain the following relation between y and s :

$$y^2 - s^2 = c^2 \tag{2.17}$$

Solving (2.17) for s^2 and carrying into the last of the relations (2.11), we write these relations as follows :

$$T_0 = wc \qquad W = ws \qquad T = wy \tag{2.18}$$

The last relation shows that the tension at any point D of the cables is proportional to the vertical distance from D to the horizontal line representing the x axis.

The span of the cable is the distance L between the supports A and B of the cable having the same elevation while sag the cable is the vertical distance h from the supports to the lowest point C. These difinitions are the same as those given in the case of parabolic cables, biut it should be noted that the sag h is now

$$h = y_A - c \tag{2.19}$$

Because of our choice of coordinates axes, it should also be observed that certain ceatenary problesm involve transcendental equations which must be solved by successive approximations. However load can be assumed uniformly distributed *along the horizontal* and the catenary can be replaced by a parabola when the cable is fairly taut. This greatly simplilfies the solution of the problem, and the error introduced is small.

The position of the lowest point of the cable is not known when the supports A and B have different elevations. The problem can be solved in a manner similar to that indicated for parabolic cables, by expressing that the cable must pass through the supports and that $x_B - x_A = L$ and $y_B - y_A = d$, where L and d denote, respectively, the horizontal and vertical distances between the two supports.

3

Trusses

In this chapter, we shall discuss three broad categories of engineering structures.

1. *Trusses,* are designed to support loads and these are usually stationary, fully constrained structures. Trusses consists of exclusively straight members connected at joints located at the end of each member. Members of truss are acted upon by two equal and opposite forces directed along the members *i.e.*, are two force members.
2. *Frames*, are also designed to support loads and are also usually stationary, fully constrained structures, frames always contain at least one *multiforce member, i.e.,* a member acted upon by three or more forces which, in general, ar not directed along the member.
3. *Machines*, are disigned to transmit and modify forces and are structures containing moving parts. Machines always contain at least one multiforce member just like frames.

TRUSSSES

Definition of a Truss

One of the major types of engineering structures is truss. It provides both a practical and an economical solutuion to may engineering situations, especially in the desing of bridges and buildings. A typical truss is shown in Figure 3.1*a*. A turss consists of straight members connected at joints. No member is continuous through a joint because members are connected at their extremities only. In Figure 3.2*a*, for example, there is no member *AB* instead there are two distinct members *AD* and *DB*. Most actual structures are made of several trusses joined together to form a space framework. Each truss may be treated as a two dimensional structure as it is designed to carry those loads which act in its plane.

Generally, the members of a truss are slender and can support little lateral load; therefore, all loads, must be applied to the various joints, and not to the members themselves. In the case of bridge truss, where a concentrated load is to be applied between two joint, or where a distributed load is to be supported by the truss, a floor system must be provided which, through the use of stringers and floor beams, transmits the load to the joints (Figure 3.2)

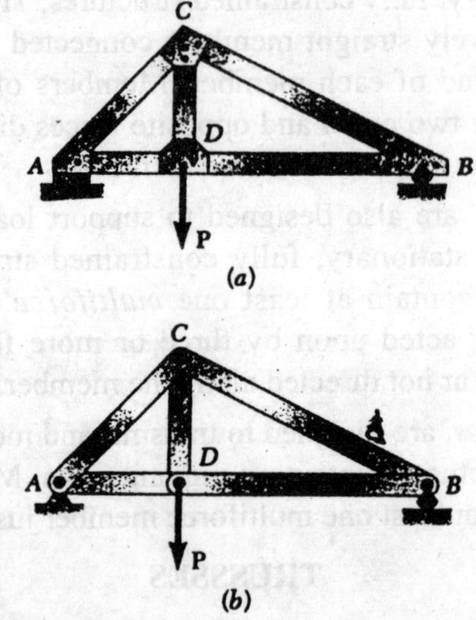

Figure 3.1

The weights of the members of the truss are supposed to be applied to the joints half of the weight of each member being applied to each of the two joints the member connects. The members are actually joined together by means of bolted or welded connections but it is customary to assume that the members ar pinnde together so that, the forces acting at each end of a member reduce to a single force and no couple. Thus, the

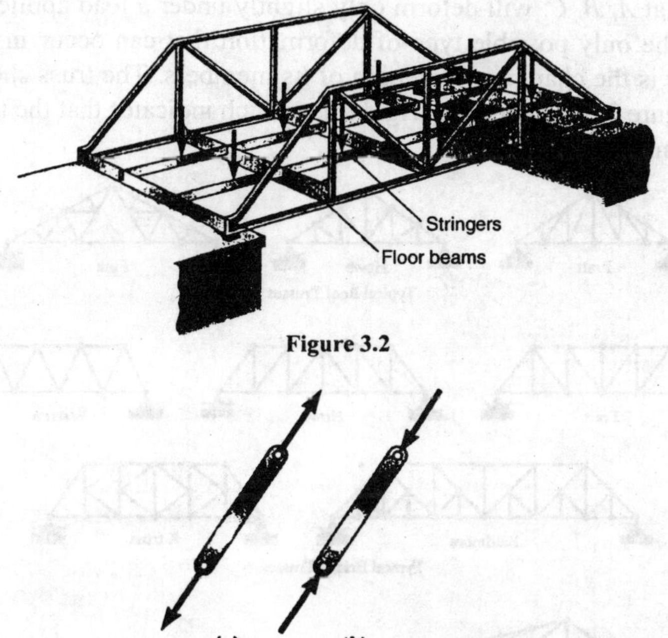

Figure 3.2

Figure 3.3

only forces assumed to be applied to a truss member are a single force at each end of the member. Entire truss can be considered as a group of pins and each member can than be treated as a two-force member (Figure 3.1*b*). An individual member can be acted upon as shown in either of two two stketches of Figure (3.3). In figure 3.3*a*, the forces tend to pull the member apart, and the member is in tension; in Figure 3.3*b*, the forces tend to compress the member, and the member is in compression. A number of typical trusses are shown in Figure 3.4.

Simple tursses

Now, let us study same simple trusses. Consider the truss of Figure 3.5*a*, which is made of four members connected by pins *A, B, C,* and *D.* If we apply a load at *B*, the truss will greatly deform and will loose its original shape completely, while the truss of Figure (3.5*b*), which is made of three members connected by pins at *A, B, C,* will deform only slightly under a load applied at *B.* The only possible type of deformation that can occur in this truss is the change in thelength of its members. The truss shown if figure 3.3*b* is known as rigid truss which indicates that the truss will not collapse.

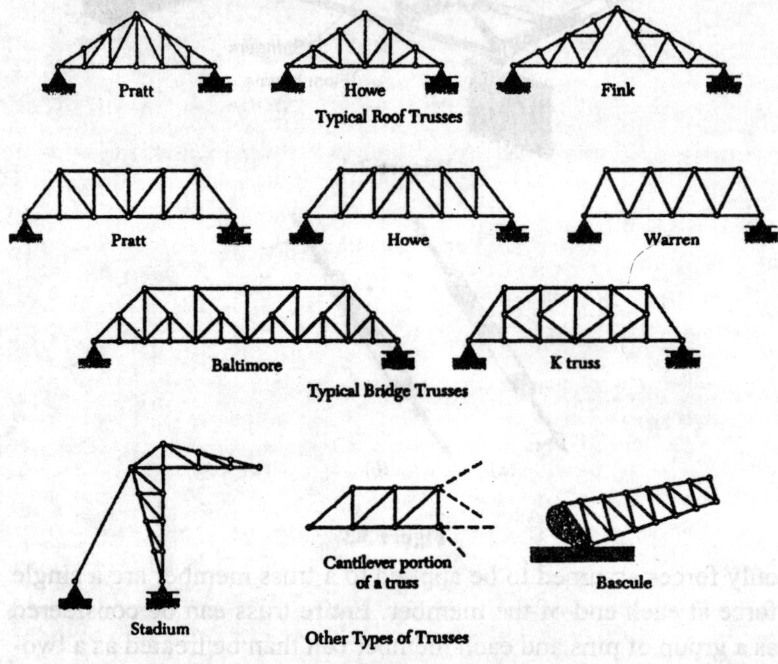

Figure 3.4

As shown in figure 3.5*c*, a larger rigid truss can be obtained by adding two members *BD* and *CD* to the basic triangular truss of Figure 3.5*b*. This procedure can be repeated as many time as desired and if each time the two members added are attached to two existing joints and then connected at a new joint, the resulting truss will be rigid. A truss which can be constructed in this manner is called a *simple truss.*

It should also be noted that a simple truss in not necessarily made only of triangles: The truss of Figure 3.5*d*, for example, is a simple truss which was constructed from triangle *ABC* by adding successively the joints *D, E, F,* and *G.* On the other hand, rigid trusses are not always simple trusses, even when they appear to be made of triangles. The Fink and Baltimore trusses shown in Figure 3.4 cannot be constructed from a single triangle in the manner described above so they are not simpe trusses. All the other trusses shown in Figure 3.4 are simple trusses, as may be easily checked. (For teh K truss, wtar with one of the central traingles).

Returning of Figure 3.5, we see that the basic triangular truss of Figure 3.5*b* has two more members and three joints. The truss of Figure 3.5*c* has two more members and one more joint, *i.e.,* five members and four joints altogether. This Clearly indicates that every time when two new members are added then the number of joints is incrcascd by onc, so we can conclude that the number of members in a simple truss can be given by $m = 2n - 3$ where n = total number of joints.

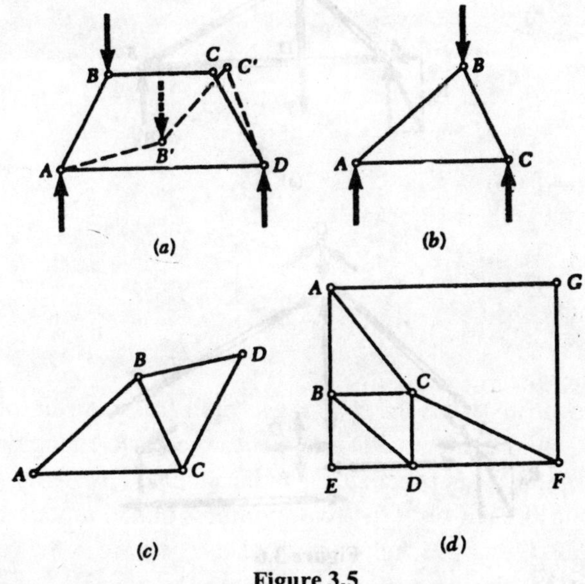

Figure 3.5

Analysis of Trusses by the method of joints

We know that a truss can be considered as a group of pins and two-force members. The truss of Figure 3.1, whose free-body diagram is shown in Figure 3.6*a*, can thus be dismembered, and a free-body diagram can be drawn for each pin and each member

(3.6*b*). Each member of a truss ia acted upon by two forces having same magnitude, same line of action but opposite sense, one at each end. Furthermore, Newton's third law indicates that the forces of action and reaction between a member and a pin are equal and opposite. Therefore, the forces exerted by a member on the two pins it connects must be directed along that member and be equal and opposite. The common magnitude of the force exerted by member of the two pins is commonly known as the force in the member considered and this quantity is scalar actually. As the line of action of all the internal forces in a truss are known, their analysis is reduced to computing the force in various members and then determining whether each member in in tension or in compression.

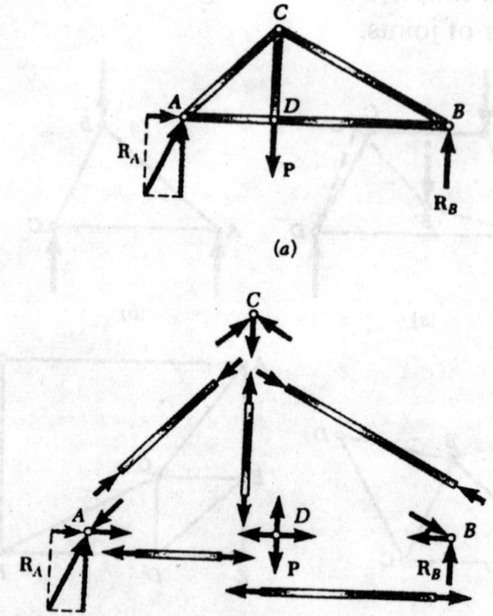

Figure 3.6

Since the entire truss is in equililbrium, each pin must be in equilibrium.This fact can be established by drawing its free-body diagram and writing two equilibrium equation. If the truss contains n pins, there will , be $2n$ equations available, which can be solved for $2n$ unknowns. For example, in the case of a simple truss, we have $m = 2n - 3$, that is, $2n = m + 3$, and the number of unknowns which can be determined from the free-body diagrams of the pins is thus $m + 3$. This means that the forces in all the members, the two components of the reaction $\mathbf{R}_A$ and the reaction $\mathbf{R}_B$ can be found by considering the free-body diagrams of the pins.

The fact that the entire truss is a rigid body in equilibrium can be used to write three more equatons involving the forces shown in the free-body diagram of Figure3.6(*a*). These equations are not independent of the equation associated with the free body diagrams of the pins as they do not contain any new information. Nevertheless, they can be used to determine the components of the reactions at the supports. The pins and members in a simple truss are so arranged that it is always possible to find a joint involving only two unknown forces.

As and exampe we analyse the truss of Figure 3.6 by considering the equilibrium of each pin successively, starting with a joint at whichonly two forces are unknown. In the truss considered, all pins are subjected to at least three unknown forces. By considering the entire truss as a free body and using the equations of equilibrium of a rigid body, firstly we determine the reaction at the support, in this way we find that $\mathbf{R}_A$ is vertical and determine the magnitudes of $\mathbf{R}_A$ and $\mathbf{R}_B$.

The number of unknown forces at joint *A* is now reduced to two, and these forces can be determined by considering the equilibrium of pin *A*. The reaction $\mathbf{R}_A$ and the forces $\mathbf{F}_{AC}$ and $\mathbf{F}_{AD}$ exerted on pin *A* by members *AC* and *AD*, respectively, must form a force triangle. Noticing the fact that $\mathbf{F}_{AC}$ and $\mathbf{F}_{AD}$ are directed along *AC* and *AD* respectively, first we draw $\mathbf{R}_A$ and then remembering the above fact we complete the triangle and detemine the magnitude and sense $\mathbf{F}_{AC}$ and $\mathbf{F}_{AD}$.. The magnitudes F_{AC} and F_{AD} represents the froces in members *AC* and *AD*. As F_{AC} is directed towards joint *A* that is down and to the left, member *AC* pushes on pin *A* and is in compression. Since $\mathbf{F}_{AD}$ is directed *away* from joint *A*, member *AD* pulls on pin *A* and is in tension.

We now proceed to joint *D*, where only two forces $\mathbf{F}_{DC}$ and $\mathbf{F}_{DB}$ are still unknown. The other forces are the given load *P*, and

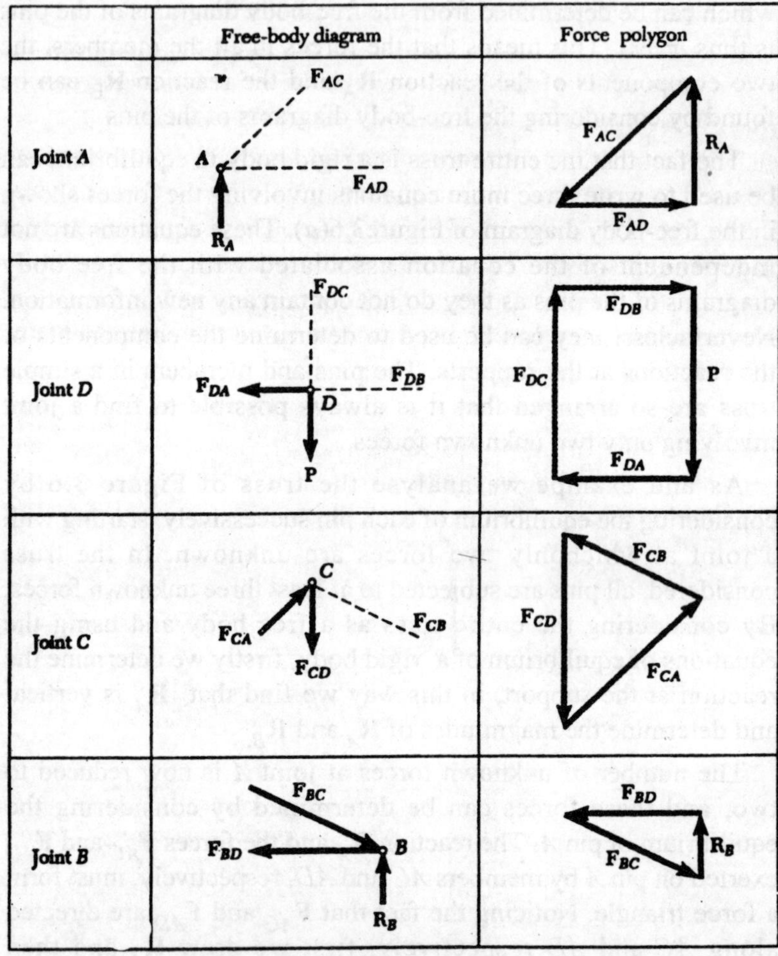

Figure 3.7

the force $\mathbf{F}_{DA}$ exerted on the pin by member *AD*. As shown above, this force is equal and oposite to the force $\mathbf{F}_{AD}$ exerted by the same member on pin *A*. The force polygon corresponding to joint *D* can be easily drawn as shown in figure 3.7 and we can determine the forces $\mathbf{F}_{DC}$ and $\mathbf{F}_{DB}$ from that polygon. It is usually more convenient to solve the equation of equilibrium $\Sigma F_x = 0$ and $\Sigma F_y = 0$ for the two unknown forces when more than three forces are involved. Since both of these forces are found to be directed away from joint

D, members *DC* and *DB* pull on the pin and are in tension.

Now, we consider the joint *C* and its free-body diagram is shown in Figure 3.7. It is noted that both $\mathbf{F}_{CD}$ *and* $\mathbf{F}_{CA}$ are known from the analysis of the preceding joints and that only $\mathbf{F}_{CB}$ is unknown. Since sufficient information to determine two unknowns is provided by the equilibrium of each pin, a check of our analysis is obtained at this joint. The force triangle is drawn, and the magnitude and sense of $\mathbf{F}_{CB}$ are determined. Since $\mathbf{F}_{CB}$ is directed toward joint *C*, member *CB* pushes on pin *C* and is in compression. By verifying that the force $\mathbf{F}_{CB}$ and member *CB* are parallel, the check is obtained.

At joint *B*, all of the forces are known. Since the corresponding pins in in equilibrium, the force triangle must close and an additional check of the analysis is obtained.

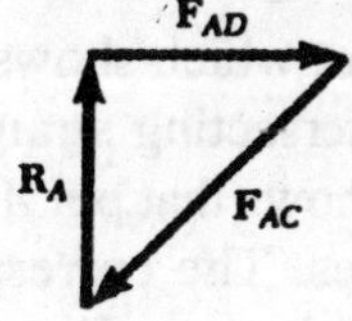

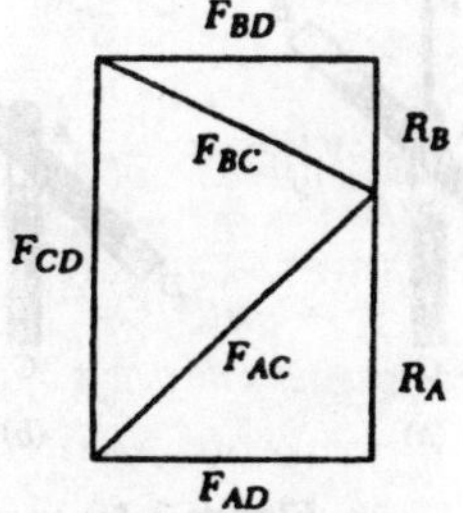

Figure 3.8

As a matter of fact, the forces polygons shown in Figure 3.7 are not unique. An alternative configuration can replace each of them. For example, the force-traingle corresponding to joint *A* could be shown in Figure 3.8. The triangles actually shown in Figure 3.7 was obtained by drawing the three forces $\mathbf{R}_A$, $\mathbf{F}_{AC}$, and $\mathbf{F}_{AD}$ in tip-to-tail fashion n the order in which their linces of action are encountered when moving clockwise around joint *A*. The other force polygons in Figure 3.7, having been drawn in the same way, can be made to fit into a single diagram, as shown in Figure 3.9. Such a diagram known as *Maxwell's diagram*, greatly facilitates the *graphical analysis* of truss problems.

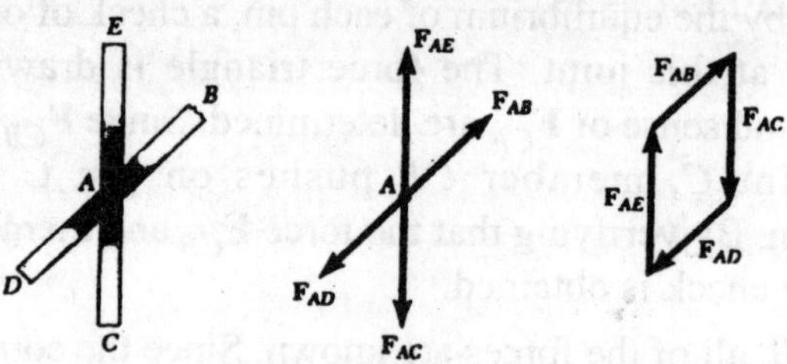

Figure 3.9

Joints under special loading conditions

Consider a Figure 3.10*a*, which shows that joint connects four members lying in two intersecting straight lines. The free-body diagram of Figure 3.10*b* shows that pin *A* is subjected to two pairs of directly opposite forces. The corresponding force polygon, therefore, must be a parallelogram (Figure 3.10*c*), and the *forces in opposite members must be equal.*

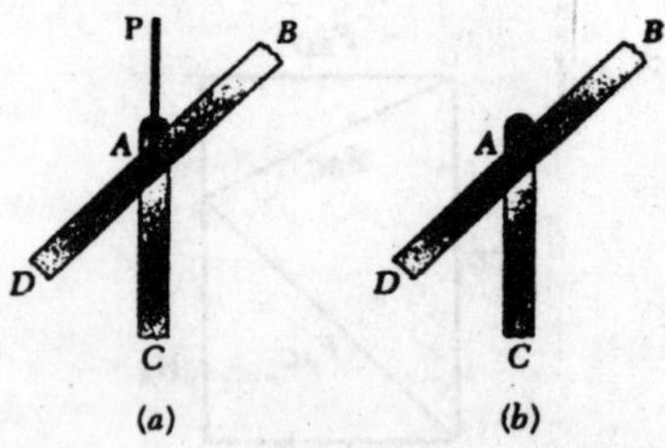

Figure 3.10

Consider next figure 3.11(*a*), in which the joint connects three members and supports a load **P**. Two of the members lie in the same lilne, and the load **P** acts along the third member. The free-body diagram of pin *A* and the corresponding force polygon will be as shown in Figure 3.10*b* and *c*, with $\mathbf{F}_{AE}$ replaced by the load **P**. Thus, *the forces in the two oppostite members must be equal, and the force in the other member must equal P.* An intersecting case is shown in Figure 3.11*b*, in this case no external load is applied to the joints, we can say $P = 0$ and also the force in member AC is zero such member having zero force is said to be a zero-force member.

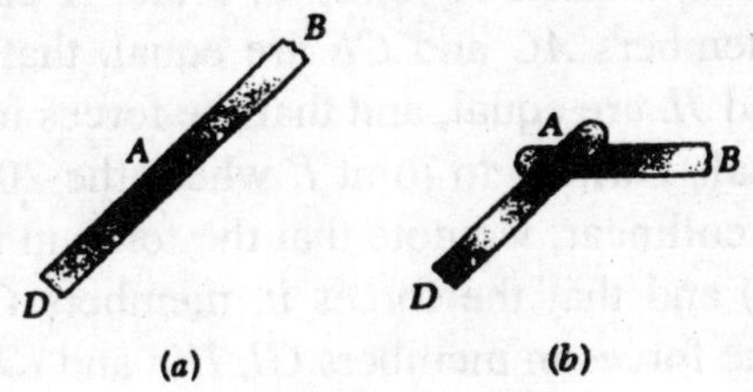

Figure 3.11

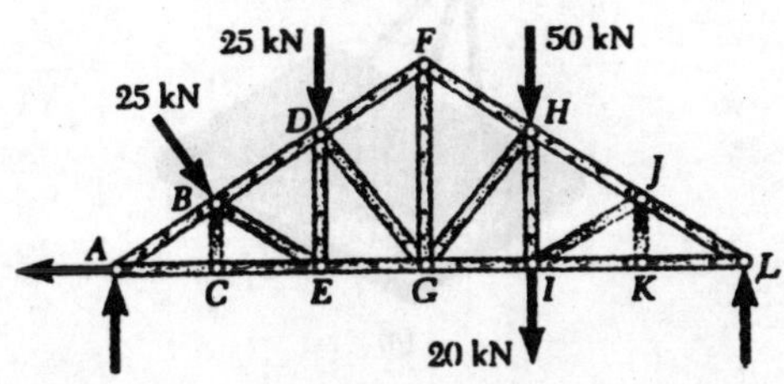

Figure 3.12

Now, consider a joint connecting two members only. We know that if the two forces acting on a particle have the same magnitude, same line of action and opposite sense than the two forces are in equilibrium. In the case of the joint of Figure 3.12*a* which connects two members *AB* and *AD* lying in the same line, *the forces in the two members must be equal* for pin *A* to be in equilibrium. In case of the joint of Figure 3.12*b*, pin *A* cannot be in equilibrium unless the forces in both members are zero. Members connected as shown in Figure 3.12*b* should essentialy be *zero-force memebers.*

Spotting the joints which are under the special loading conditions listed above will expedite the analysis of a truss, for example, a Howe truss loaded as shown if Figure 3.13. Zero force member are represented by green lines. Joint *C* connects three members, two of which lie in the same line, and is not subjected to any external load; member *BC* is thus a zero-force member.

Now let us apply same reason to joint *K*, we find that member *JK* is also a zero-force member. But joint *J* is now in the saem situation as joint *C* and *K*, and member *J* must be a zero-force member. The examination of joints *C, J* and *K* also shows that the forces in members *AC* and *CE* are equal, that the forces in members *HJ* and *JL* are equal, and that the forces in members *IK* and *KL* are equal. Turning to joint *I*, where the 20-kN load and member *HI* are collinear, we note that the force in member *HI* is 20 kN (tension) and that the forces in members *GI* and *IK* are equal. Hence, the forces in members *GI, IK,* and *KL* are equal.

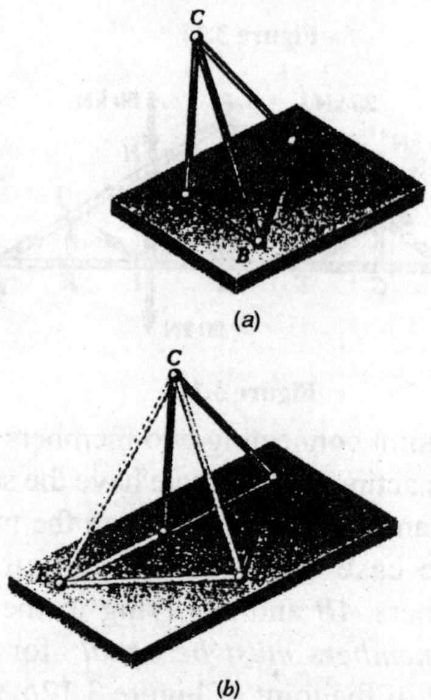

Figure 3.13

Note that the condition described above are not applicable to joints *B* and *D* in Figure 3.13 and also it would be fairly wrong to suppose that the force in member *DE* is 25 kN or the memberr *AB* and *BD* have equal forces. By carrying out the analysis of joints *A, B, D , E, F, G ,H* and *L* in the usual manner the forces in thses members and all remaining members be found. Thus until you have become throughly familiar with the conditions under which the rules established in this section can be applied, you should draw the free-body diagrams of all pins and write the corresponding equilibrium equations (or draw the corresponding force polygons) whether or not the joints being considered are under one of the special loading conditions described above

Importance of zero force members

These are not entirely useless for instance in the figure 3.18 the zero force member do not carry any load under the shown loading condition but the same member would propable be required to carry loads if loading conditions were subjected to change. Even in the case considered zero force members are needed to maintain the desired shape of the truss and also to support the weight of the truss.

Space trusses

A special truss is obtained when several straight members are joined together at their extremities to form a three dimensional configuration.

We recall from Section 3.2 that the most elementary two-dimensional rigid truss consisted of three members joined at therir extremities to form the sides of a triangle; by adding two members at a time to theis basic configuration, and connecting them at a new point, it was possible to obtain a larger rigid structure which was called a simple truss. In the same way the most elementary rigid space truss consists of six members joined at their extremities to form the edges of a tetrahydron *ABCD* (figure 3.14*a*). By adding three members at a time to this basic configuration, such as *AE, BE,* and *CE*, attaching them to three existing joints, and connecting them at a new joint , (all the four joint must not lie in a plane) we can obtain a larger rigid structure which is defined as a *simple space trus* (Figure 3.14*b*). Observing that the basic tetrahedron has six members and for joints and that every time three members are added, the number of joints is increased by one, we conclude that in a simple space truss the total number of members is $m = 3n - 6$, where n is the total number of joints.

If a space truss is to be completely constrained and if the reaction at its supports are to be statically determinate, the supports should consist of a combination of balls, rollers, and balls and sockets which provides six unknown reactions. By expressing that the three dimensional truss is in equililbrium, these unknown reaction may be readily determined by solving the six equation.

In reality the members of a space truss are actually joined together by means of bolted or welded connections but it is supposed that each joint consists of a ball-and-socket connection. Thus, no couple will be applied to the members of the truss, and each member can be treated as a two-force member. The three equation $\Sigma F_x = 0$, $\Sigma F_y = 0$, and $\Sigma F_z = 0$ express the condition of equilibrium for each joint. In the case of a simple space truss having n joints, writing the conditions of equilibrium for each joint will thus yield $3n$ equations. Since $m = 3n - 6$, these equations suffice to determine all unknown forces (forces in m members and six reactions at the supports). We should select joints in such an order that no selected joint involve more than three unknown forces to avoid the necessity of solving simultaneous equations.

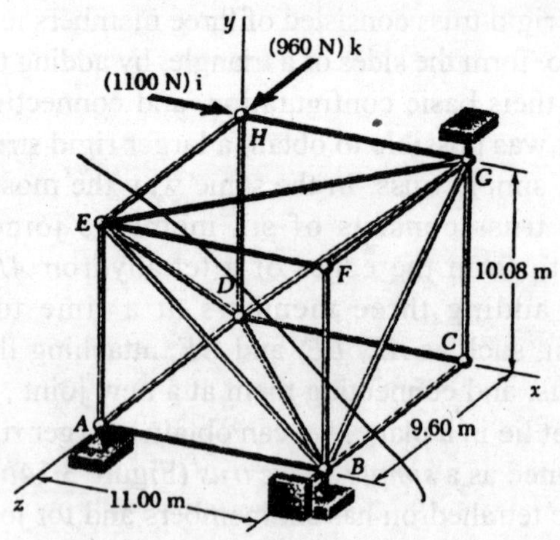

Figure 3.14

Analysis of trusses by the method of sections

When the forces in all the members of a truss are to be determined then the method of joints is the most effiecient. But when the force in one member or the force in very few member is required then the method of section is more effective.

For example, if we want to determine the force in member *BD* of the truss shown in Figure 3.15 (*a*), we must determine the force with which member *BD* acts on either joint *B* or joint *D*. If we use the method of joints, we choose either joint *B* or joint *D* as a free-body. However, we can also choose a larger portion of the truss, composed of several joints and members as a free body, provided that the desired force is one of the external force acting on that portion. If in addition, the portion of truss which is chosen has a total of only three unknown forces acting upon it, the desired force can be obtained by solving the equations of equilibrium for this portion of the truss. By passing a section through three members of the truss, one of which is the desired member, the portion of the truss to be ulilized is obtained *i.e.* by drawing a line which divides the truss into two completely separate parts but does not intersect more than three members. Either of the two portions of the truss obtained can be used as a free body, after the intersected membes have been removed.

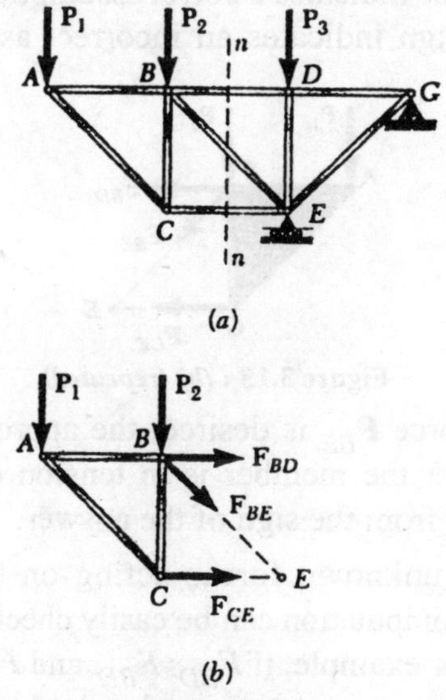

Figure 3.15

In Figure 3.15*a*, the section *nn* has been passed through members *BD, BE,* and *CE*, and the portion *ABC* of the truss is chosen as the free body. The forces acting on the free body are the loads $\mathbf{P}_1$ and $\mathbf{P}_2$ at points A and B and three unknown $\mathbf{F}_{BD,}$ $\mathbf{F}_{BE,}$ and $\mathbf{F}_{CE}$. The three forces have been arbitrarily drawn away from the free body as if the members were in tension, as we do not know whether the members were in tension or compression.

The fact that the rigid body ABC is in equilibrium can be expressed by writing three equations which can be solved for the three unknown forces. If only the forces $\mathbf{F}_{BD}$ is desired, we need write only one equation, provided that the equation does not contain the other unknowns. Thus the equation ΣM_E yields the value of the magnitude F_{BD} of the forces $\mathbf{F}_{BD}$ (Figure 3.15*b*). A positive sign in the answer implies that our assumption regarding the sense of F_{BD} was correct and the member BD is in tension whereas negative sign indicates that our assumption regarding F_{BD} was wrong and the member is in compression.

On the other hand, if only the force $\mathbf{F}_{CE}$ is desired, an equation which does not involve $\mathbf{F}_{BD}$ or $\mathbf{F}_{BE}$ is written; the appropriate equation is $\Sigma M_B = 0$. Again a positive sign for the magnitude F_{CE} of the desired force indicates a corret assumption, that is, tension; and a negative sign indicates an incorrect assumption, that is, compression.

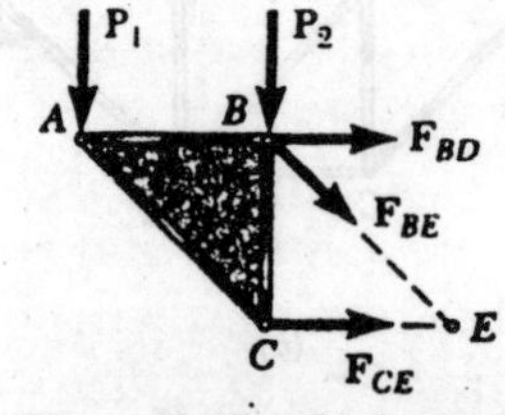

Figure 3.15 : *(b) (repeated)*

In only the force $\mathbf{F}_{BE}$ is desired, the appropriate equation is $\Sigma F_y = 0$. Whether the member is in tension or compression is again determined from the sign of the answer.

When all the unknown forces acting on the free body are determined, the computation can be easily checked by writing one more equation, for example, if F_{BD}, F_{BE}, and F_{CE} are determined as shown above the computation can be checked by verifying that $\Sigma F_x = 0$. However if the force is only one member is determined, ther is no independent check of the computation.

Trusses made of several simpe trusses

Consider two simple trusses *ABC* and *DEF*. If they are connected by three bars *BD, BE,* and *CE* as shown in Figure 3.16*a*, they will form together a rigid truss *ABCD*. The trusses *ABC* and *DEF* can also be combined into a simgle rigid truss by joining *B* and *D* into a single joint *B* and by connecting *C* and *E* by a bar *CE*. (Figure 2.16*b*). The truss thus obtained is known as a *Fink truss*. It should be very clear that the trusse of Figure 3.16*a* and *b* are not simple trusses as they cannot be prepared from a triangular truss by adding successive pairs of members as described in previous section. They are rigid trusses. Trusses made of several simple trusses rigidly connected are known as compound trusses.

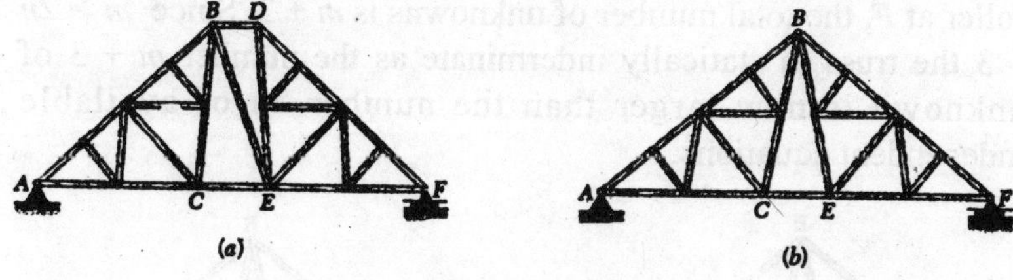

Figure 3.16

Also, in a compound truss the number of members m and the number of joints n are related by same formula $m = 2n - 3$. This can be checked by observing that, if a compound truss is supported by a frictionless pin and a roller (involving three unknown reacitons), the total number of unknowns is $m + 3$, and this number must be equal to the number $2n$ of equations obtained by expressing that the n pins are in equilibrium; it follows that $m = 2n - 3$. Compound trusses are statically determinate, rigid and completely constrained if they are supported by a pin and a roller or by an equivalent system of supports. This means that all of the unknown reactions and the forces in all the members can be determined by the methods of statics, and that the truss will neither collapse nor move. The forces in the members, however, cannot all the determined by the method of joints, except by solving a large number of simultaneous equations. In the case of the compooud truss of Figure 3.16*a*, for example to determine the forces in these members it is more sufficient to pass a reaction through member *BD, BE,* and *CE*.

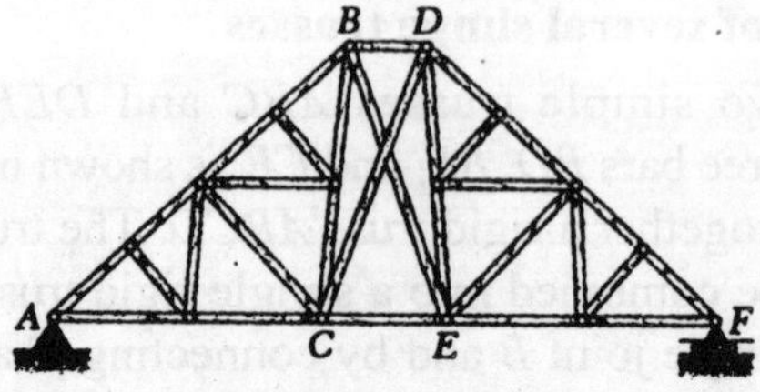

Figure 3.17

Suppose, now that the simple trusses *ABC* and *DEF* are connceted by *four* bars *BD, BE, CD,* or *CE* (Figure 3.17). The number of members *m* is now larger than $2n - 3$; the truss obtained is overrigid and one of the four members *BD*, *BE*, *CD* or *CE* is said to be *redundant.* If the truss is supported by a pin at *A* and a roller at *F*, the total number of unknowns is $m + 3$. Since $m > 2n - 3$ the truss in statically inderminate as the number $m + 3$ of unknown is now larger than the number $2n$ of available independent equations.

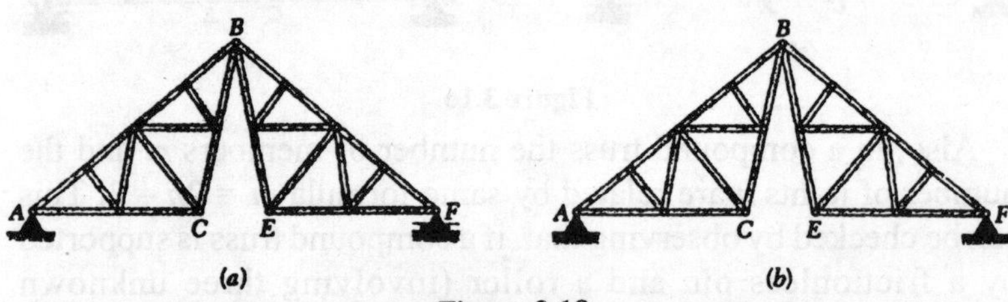

Figure 3.18

Finally, let us assume that the two simple trusses *ABC* and *DEF* are joined by a pin as shown in Figure 3.18*a*. The number of members *m* is smaller than $2n - 3$. The total number of unknown is $m + 3$ when the truss is supported by a pin at *A* and a roller at *F*. Since $m < 2n - 3$, the number $m + 3$ of unknowns is now smaller than the number $2n$ of equilibrium equations which should be safisfied; the truss in *nonrigid* and will collapse under its own weight. In order to get a rigid truss that will not collapse (Figure 3.18*b*) two pins are used to supposed it. We note that the total number of unknowns in now $m + 4$ and is equal to the nuber $2n$ of equations. More generally, if the reactions at the supports involve *r* unknowns, the condition for a compound truss to be statically determinate, rigid, and completely constrained is $m + r = 2n$. However, this condition is not sufficient for the equilibrium of a structure which ceases to be rigid when detached from its supports.

4

Moments and Products of Inertia

Introduction

In this chapter, we shall discuss certain measres of mass distribution relative to a reference which are vital for the study of the dynamics of rigid bodies. These quantities are so closely related to second moments and products of area, we shall consider them at this early stage rather than wait for dynamics. The fact that these measures of mass distribution- the second moments of inertia of mass and the products of inertia of mass-are components of second-order tensor, shall also be discussed. By recognising this fact early we can make our further studeis of stress and strain more simple clear and understandable. Since these quantities also happen to be componentsw of second-order tensors.

Formal definition of Inertia Quantities

Now we shall formally define a set of quantities that give information about the distribution of mass of a body relative to a

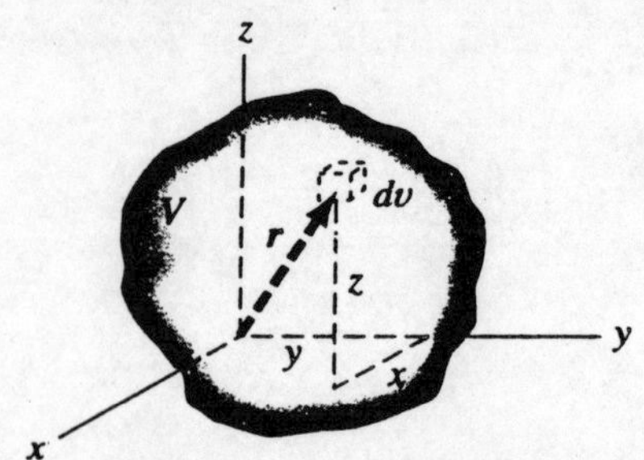

Figure 4.1 : ***Body and reference at time*** t.

Cartesian reference. For this purpose, Figure 4.1 shows a body of mass *M* and a reference *xyz*. This referece and the body may have any motion whatever relative to each other. The ensuring discussion then holds for the instantaneous orientation shown at time *t*. We shall assume that the body is composed of a continuum of particles, eachof which has a mass give by *p dv*. Now the following definitions are presented.

$$I_{xx} = \iiint\limits_V \left(y^2 + z^2\right) \rho \, dv \tag{4.1a}$$

$$I_{yy} = \iiint\limits_V \left(x^2 + z^2\right) \rho \, dv \tag{4.14}$$

$$I_{zz} = \iiint\limits_V \left(x^2 + y^2\right) \rho \, dv \tag{4.1c}$$

$$I_{xy} = \iiint\limits_V xy \, \rho \, dv \tag{4.1d}$$

$$I_{xz} = \iiint\limits_V xz \, \rho \, dv \tag{4.1e}$$

$$I_{yz} = \iiint\limits_V yz \, \rho \, dv \tag{4.1f}$$

The *mass moments of inertia* of the body about the *x, y,* and *z* axes, are represented by term $I_{xx,}$ I_{yy} and I_{zz} respectively. Note that in each such case we are integratiing the mass elements ρ *dv*, times the *perpendicular distance squared* from the mass elements to the coordinate axis about which we are computing the moment of inertia. Thus, if we look along the *x* axis toward the origin in Figure 4.1, we would have the view shown in Figure 4.2. The quantity $y_2 + z_2$ used in Equation 4.1a for I_{xx} is clearly d^2 the perpendicular distance squared from *dv* to the *x* axis (now seen as a dot). Each of the terms with mixed indices is called the *mass product of inertia* about the pair of axes given by the indices. Clearly, from the definition of the product of inertia, we could reverse indices and thereby form three additional products of inertia for a reference. The additional three quantities formed in this way, however, are equal to the corresponding quantities of the original set. That is,

$$I_{xy} = I_{yx}, I_{xz} = I_{zx}, I_{yz} = I_{yz}$$

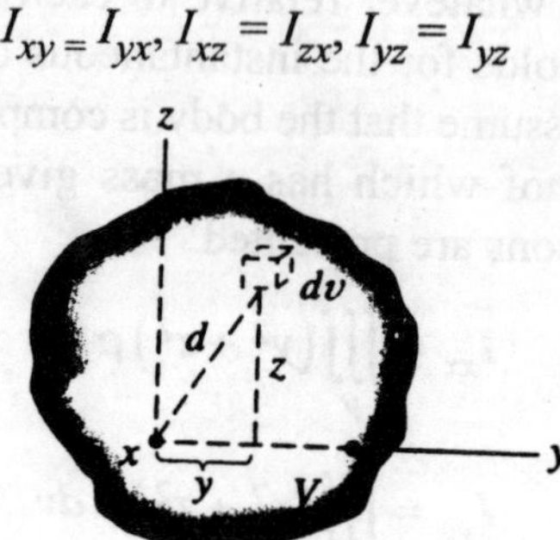

Figure 4.2 : ***View of body along x axis.***

We have nine inertia terms at a point for a given reference at this point. For a given body, the value of the set of six independent quantities will, depend on the position and *inclination* of the reference relative to the body. The refernce need not be situated in the rigid body of interest and can by established anywhere in space.Thus there will be nine inertia terms for reference *xyz* at point *O* outside the body (Fig. 4.3) computed usings Equations 4.1, where the domain of integration is the volume *V* of the body. The nine moments and products of inertia are components of the inertia tensor, this will be explained later.

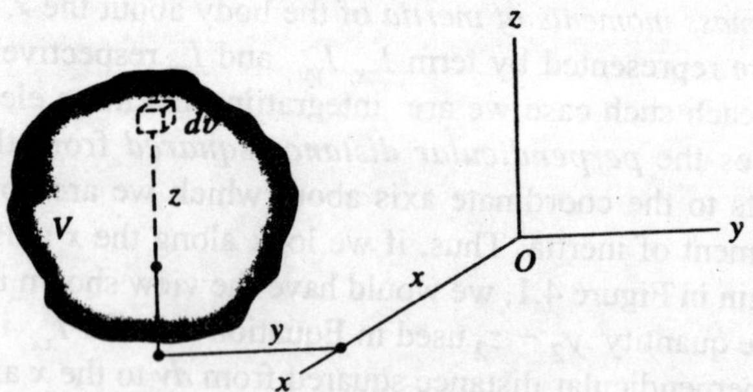

Figure 4.3 : ***Origin or* xyz *outside body***

The nine moments and products of inertia for reference *xyz* at a point can be made more convinent for study by listing them in a matrix array as follows :-

$$I_{ij} = \begin{pmatrix} I_{xx} & I_{xy} & I_{xz} \\ I_{yx} & I_{yy} & I_{yz} \\ I_{zx} & I_{zy} & I_{zz} \end{pmatrix}$$

Notice that the row gives the first subscript and the second subscript gives the column in the array. Furthermore, the mass moment of inertia terms is composed by the left to right downward diagonal in the array. while the products of inertia, oriented at mirror-image positions about this diagonal, are equal. The array is *symmetric* for this reason.

The sum of the mass moments of inertial for a set of orthognal axes depends only on the position of the origin & is independent of the orientation of the axes.

$$I_{xx} + I_{yy} + I_{zz} = \iiint_V (y^2 + z^2)\rho\, dv + \iiint_V (x^2 + z^2)\rho\, dv + \iiint_V (x^2 + y^2)\rho\, dv$$

On rearranging and combining integrals we obtain :-

$$I_{xx} + I_{yy} + I_{zz} = \iiint_V 2(x^2 + y^2 + z^2)\rho\, dv = \iiint_V 2|r|^2 \rho\, dv \qquad 4.2$$

But the magnitude of the position vector from the origin to a particle is *independent* of the inclination of the reference at the origin.. For a given body, the sum of the moment of inertia at a point in space is an invariant with respect to rotation of axes.

Clearly on examining equation 4.1 we know that the moments of inertia are always positive, while the product of inertia may be positive or negative. Of interest is the case where one of the coordinate planes is a plane of *symmetry* for the mass distribution of the body. Such a plane is the *zy* plane shown in Figure 4.4 cutting a body into two parts, which, by definition of symmetry, are mirrors images of each other. For the computation of I_{xz} each half will give a contribution of the same magnitude but of opposite sign. We can most readily see that this is so by looking along the *y* axis towards the origin. Figure 4.5 shows the plane of symmetry that appear as a line coinciding with the *z* axis.. We can consider

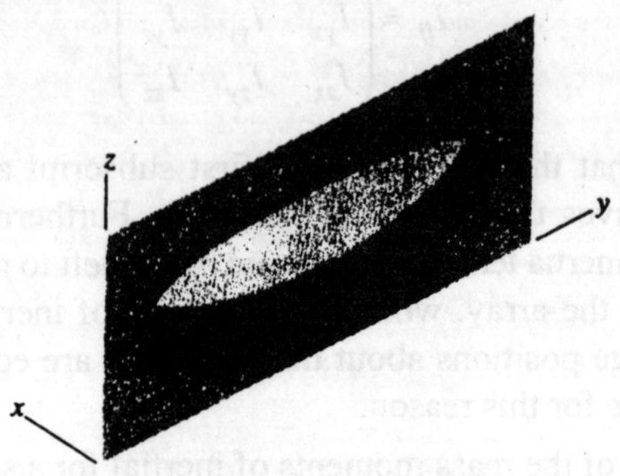

Figure 4.4 : *zy is plane of symmetry*

the body to be composed of pairs of mass elements *dm* which are mirrors images of each other with respect to position and shape about the plane of symmetry. For such a pair the product of inertia I_{xz} is then :

$$xz\,dm - xz\,dm = 0$$

Thus, we can conclude that

$$I_{xz} = \underbrace{\int xz\,dm}_{\text{right domain}} - \underbrace{\int xz\,dm}_{\text{left domain}} = 0$$

Figure 4.5 : *View along y axis*

For I_{xy} also, this conclusion is true. We can say that $I_{xy} = I_{xz} = 0$. But on consulting Figure 4.4, you should be able to readily decide that the term I_{zy} will have a positive value. Note that those products of inertria having x as an index are zero and that the x coordinate axis is normal to the plane of symmetry. Thus, we can conclude that *if two axes form a plane of symmetry for the mass distribution of a body, the products of inertia having as an index the coordinate that is normal to the plane of symmetry will be zero.*

Consider next a body of *revolution*. Take the z axis to coincide with the axis of symmetry. It is easy to conclude for thr origin O of xyz anywhere along the axis of symmetry that

$$I_{xz} = I_{zy} = I_{xy} = 0$$
$$I_{xx} = I_{yy} = \text{constant}$$

for all possible xy axes formed by rotating about the z axis at O. Can you justify these conclusions .

Finally, we define *radii of gyration* as ;

$$I_{xx} = k^2_x M$$
$$I_{yy} = k^2_y M$$
$$I_{zz} = k^2_z M$$

where k_x, k_y and k_z are the radii of gyration and M is the total mass.

Relation between Mass-inertia terms and area-inertia terms

Now we relate the second moments and product of area with the inertia tensor. For this purpose, consider a plate of constant thickness t and uniform density ρ (Fig 4.6). A reference is selected so that the xy plane is the midplane of this plate. For convenience the components of the inertia tensor are rewritten as :

$$I_{xx} = \rho \iiint_V (y^2 + z^2)\, dv, \qquad I_{xy} = \rho \iiint_V xy\, dv$$

$$I_{yy} = \rho \iiint_V (x^2 + z^2)\, dv, \qquad I_{xz} = \rho \iiint_V xz\, dv \qquad 4.3$$

$$I_{zz} = \rho \iiint_V (x^2 + y^2)\, dv, \qquad I_{yz} = \rho \iiint_V yz\, dv$$

Now assume that the thickness t is *small* compared to the lateral dimension of the plate, which restrict the z to a range of values

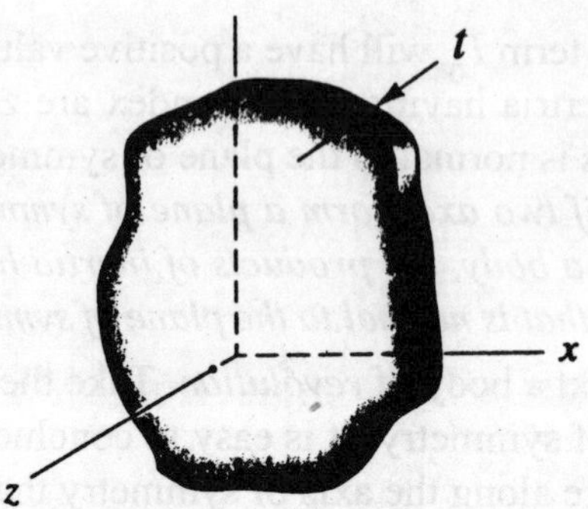

Figure 4.6 : *Plate of thickness t*

having a small magnitude. As a result, we can make two simplifications in the equations above. First whenever z appears on the right side of the equation above we shall set it (z) equal to zero. Second, we shall express dv as

$$dv = t\,dA$$

where dA is an area element on the *surface* of the plate, as shown in Figure 4.7, then equation 4.3 becomes :

$$I_{xx} = \rho t \iint_A y^2\,dA, \qquad I_{xy} = \rho t \iint_A xy\,dA$$

$$I_{yy} = \rho t \iint_A x^2\,dA, \qquad I_{xz} = 0$$

$$I_{zz} = \rho t \iint_A \left(x^2 + y^2\right) dA, \qquad I_{yz} = 0$$

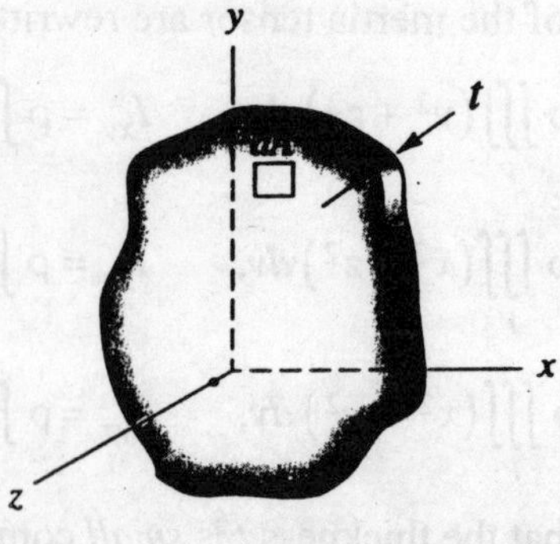

Figure 4.7 : *Use volume elements d dA.*

Now note that the integrals on the right sides of the equations above are moments and products of *area*. Denoting mass--moment and product of inertia terms with a subscript M and second moment and product of area terms with a subscript A, we can then say for the nonzero expressions :

$$(I_{xx})_M = \rho t (I_{xx})_A$$

$$(I_{yy})_M = \rho t (I_{yy})_A$$

$$(I_{zz})_M = \rho t (J)_A$$

$$(I_{xy})_M = \rho t (I_{xy})_A$$

We can compute the inertia tensor components for reference xyz for a thin plate with a constant product ρt throughout (seeFig 4.6) by using the second moments and product of area of the surface of the plate relative to axes xy.

It is important to point out that ρt is the *mass per unit area* of the plate. Imagine next that t goes to zero and simultaniously ρ goes to infinity at rates such that the product ρt becomes unity in the limit. One might thing of the relulting body to be a *plane area*. We have thus formed a plane area form a plate by our approach and in this way we can think of a plane area as a special mass. This explains why we use the same notation for mass moments and products of inertia as we use for second moments and products of area. However the units clearly will be different.

Translation of Coordinate Axes

We will compute mass moment and product of inertia quantities for a reference xyz in this section, that is displaced under a translation (no rotation) from a reference $x'y'z'$ at the centre of mass (Figure 4.8) for which the inertia terms are presumed known. Let us first compute the mass of moment of inertia I_{zz}.

Observing Figure 4.8, we see that

$$\boldsymbol{r} = \boldsymbol{r}_c + \boldsymbol{r}'$$

Hence,

$$x = x_c + x'$$
$$y = y_c = x'$$
$$z = z_c + z'$$

Now I_{zz} is formulated in the following way:

$$I_{zz} = \iiint_V (x^2 + y^2)\,\rho\, dv = \iiint_V \left[(x_c + x')^2 + (y_c + y')^2\right] \rho\, dv \qquad (4.4)$$

on squarring and rearranging we have :

$$I_{zz} = \iiint_V (x_c^2 + y_c^2)\,\rho\, dv + 2\iiint_V x_c x' \rho\, dv$$
$$+ 2\iiint_V y_c y'\,\rho\, dv + \iiint_V (x'^2 + y'^2)\,\rho\, dv \qquad (4.5)$$

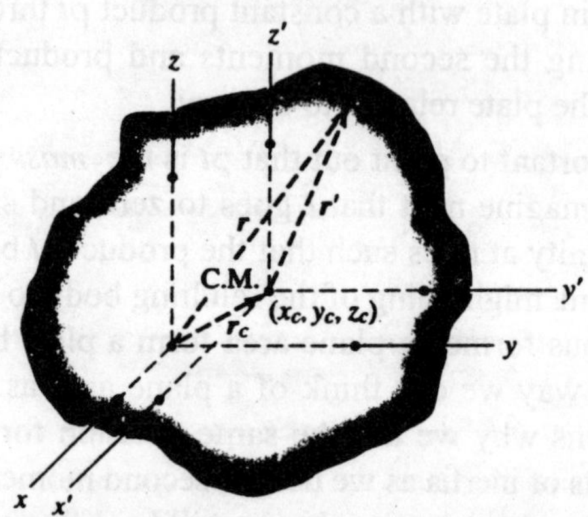

Figure 4.8 : xyz *translated from x'y'z' at C.M.*

the quantities bearing the subscript c are constant for the integration and therefore can be extracted from under the integral sign. Thus, the equation brcomes

$$I_{zz} = M(x_c^2 + y_c^2) + 2x_c \iiint_V x'\, dm$$
$$+ 2y_c \iiint_V y'\, dm + \iiint_V (x'^2 + y'^2)\,\rho\, dv \quad (4.6)$$

where dm replaces ρdv and the integration $\iiint_V \rho\, dv$ in the first integral has been evaluated as M, the total mass of the body. The origin of the primed reference being at the center of mass requires of the first moment of mass that $\iiint x'dm = \iiint y'dm = \iiint z'dm = 0$. The middle two terms accordingly drop out of the expression above, and we recognize the last expression to be $I_{z'z'}$. Thus the desired relation is

$$I_{zz} = I_{z'z'} + M(x^2_c + y^2_c) \tag{4.7}$$

By observing the body in Figure (4.8) along the z and z' axes (i.e. from directly above) we get a view as is shown in Figure 4.9. From this diagram, we can see that $y^2_c + x^2_c = d^2$ where d is the perpendicular distance between the z' axis through the center of mass and the z axis about which we are taking moments of inertia. We may then give the result above as

$$I_{zz} = I_{z'z'} + Md^2 \tag{4.8}$$

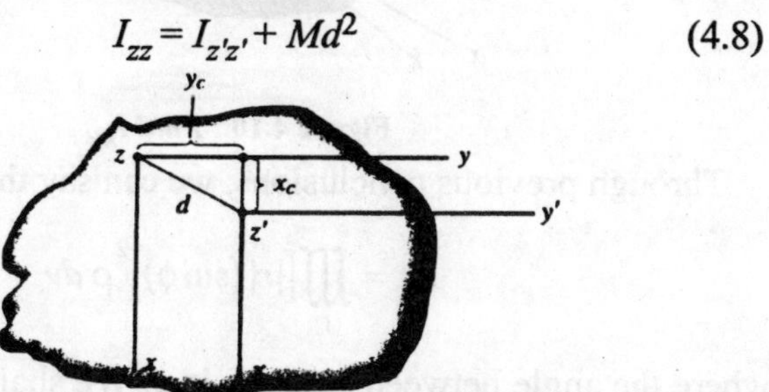

Figure 4.9 : *View along z direction (from above).*

Let us generalize from the previous statement. The moment of inertia of a body about any axis equals the moments of inertia of the body about a parallel axis that goes through the center of mass, plus the total mass times the perpendicular distance between the axes squared.

Now you have to show that for products of inertia a similar relation can be reached. For example, we have

$$I_{xy} = Ix'y' + Mx_c + Mx_c y_c \tag{4.9}$$

Here, we must take care to put in the proper signs of x_c and y_c as measured from the *xyz* reference. Equations 4.8 and 4.9 comprise the well-known *paraller axis theorems*. These equation can be used to advantage for bodies having simple familiar shapes.

Transformation properties of the Inertia Terms

Let us now suppose that the six independent inertia terms are known at the origin of a given reference. What is the mass moment of inertia for an axis going through the origin of the reference and having the directions cosines *l, m, and n,* relative to the axes of this reference. In fig. 4.10 *kk* represents the axis about which we are interested in obtaining the mass moment of inertial.

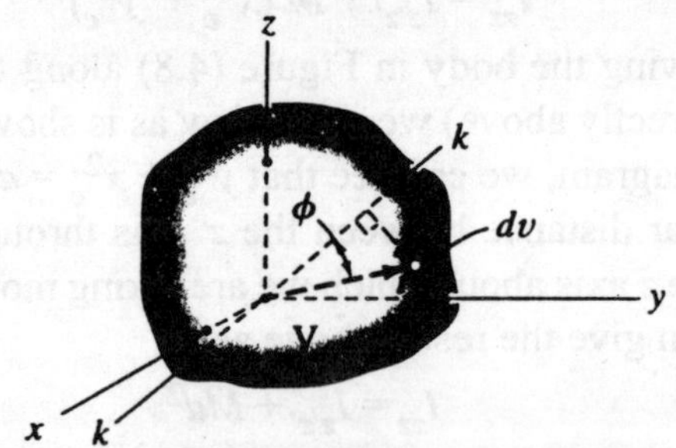

Figure 4.10 : ***Find*** I_{kk}

Through previous conclusions, we can say that

$$I_{kk} = \iiint_V \left[|r|(\sin\phi)\right]^2 \rho\, dv \tag{4.10}$$

where the angle between *kk* and *r* is ϕ. We shall put $\sin^2\phi$ into a more useful form by considering the right triangle formed by the position vector *r* and the axis *kk*. Fig. 4.11 shows the enlarged view of this triangle. The side *a* of the triangle has a magnitude that can be given by the dot product of *r* and the unit vector ε_k along *kk*. Thus,

$$a = \mathbf{r} \bullet \varepsilon_k = (x\mathbf{i} + y\mathbf{j} + z\mathbf{k}) \bullet (l\mathbf{i} + m\mathbf{j} + n\mathbf{k}) \tag{4.11}$$

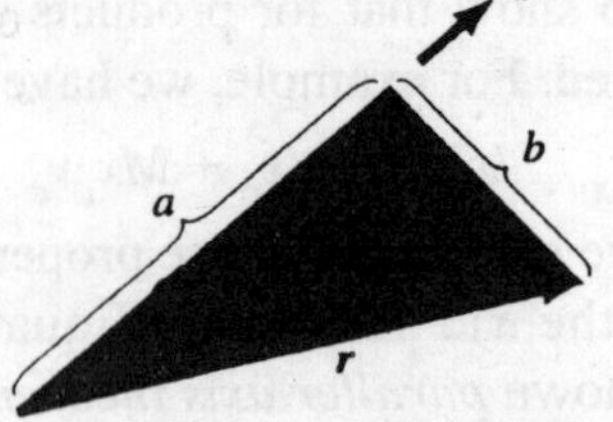

Figure 4.11 : ***Right triangle formed by r and kk.***

Hence

$$a = lx + my + nz$$

Now by using the pythagoream theorem, side b is given as;

$$b^2 = |\mathbf{r}|^2 - a^2 = (x^2 + y^2 + z^2) - (l^2x^2 + m^2y^2 + n^2z^2 + 2lmxy + 2lnxz + 2mnyz)$$

The term $\sin^2 \phi$ may be given as

$$\sin^2 \phi = \frac{b^2}{r^2} = \frac{(x^2 + y^2 + z^2) - \begin{pmatrix} l^2x^2 + m^2y^2 + \\ n^2z^2 + 2lmxy + 2lnxz + 2mnyz \end{pmatrix}}{x^2 + y^2 + z^2} \tag{4.12}$$

On, substituting back into Equation 4.10 and canceling terms, we get

$$I_{kk} = \iiint_V \left[(x^2 + y^2 + z^2) - (l^2x^2 + m^2y^2 + n^2z^2 + 2lmxy + 2lnxz + 2mnyz) \right] \rho \, dv$$

We can multiply the first bracketed exprssion in the integral by $l^2 + m^2 + n^2$ as it is equal to 1

$$I_{kk} = \iiint_V \left[(x^2 + y^2 + z^2)(l^2 + m^2 + n^2) - (l^2x^2 + m^2y^2 + n^2z^2 + 2lmxy + 2lnxz + 2mnyz) \right] \rho \, dv$$

On multiplication and collection of terms we get the relation

$$I_{kk} = l^2 \iiint_V (y^2 + z^2) \rho \, dv + m^2 \iiint_V (x^2 + z^2) \rho \, dv + n^2 \iiint_V (x^2 + y^2) \rho \, dv - 2lm \iiint_V (xy) \rho \, dv - 2ln \iiint_V (xz) \rho \, dv - 2mn \iiint_V (yz) \rho \, dv$$

We reach the desired transformation equation by referring back to the definiton presented by Equation 4.1;

$$I_{kk} = l^2 I_{xx} + m^2 I_{xy} + n^2 I_{zz} - 2lmI_{xy} - 2lnI_{xz} - 2mnI_{yz} \quad (4.13)$$

Now, we put this in a more useful form which will br of great help and use in later course in mechanics. l is the direction cosine between the k axis and the x axis. It is common practice to identify this cosine as a_{kk} instead of l. The axes involved are identified by the subscript . Similarly, m = aky and $n = a_{kz}$. We can now express Equation 4.13 in a form similar to a matrix array as follows on noting that $I_{xy} = I_{yx}$, etc.

$$\begin{aligned} I_{kk} &= I_{xx} a^2{}_{kk} - I_{xy} a_{kk} a_{ky} - I_{xz} a_{kx} a_{kz} \\ &- I_{yx} a_{ky} akx + I_{yy} a^2{}_{ky} - I_{yz} a_{ky} a_{kz} \\ &- I_{zx} a_{kz} a_{kx} - Izy_a kz_a ky + I_{zz} a^2{}_{jz} \end{aligned} \quad (4.14)$$

By first writing the matrix array of I's on the right side and then inserting the a's remembering to insert minus signs for off-diagonal terms, the above format is easily written.

Let us next compute the product of inertia for a pair of mutually perpendicular axes, Ok and Oq, as shown in Figure 4.12. The direction cosines of Ok we shall take as l, m, and n, whereas the dirction cosines of Oq we shall take as l', m', and n'. As the axes are at right angles to each other, we know that

$$\varepsilon_k \bullet \varepsilon_q = 0$$

Therefore

$$ll' + mm' + nm' = 0 \quad (4.15)$$

$r \bullet \varepsilon_k$ and $r \bullet \varepsilon_q$ are the co-ordinates of the mass elements ρdv along the axes Ok and Oq respectively. We have, for I_{kq}:

$$I_{kq} = \iiint_V (\boldsymbol{r} \bullet \varepsilon_k)(\boldsymbol{r} \bullet \varepsilon_q)\, \rho\, dv$$

Using xyz components of r and the unit vectors, we can write :

$$\begin{aligned} I_{kq} &= \iiint_V [(x\boldsymbol{i} + y\boldsymbol{j} + z\boldsymbol{k}) \bullet (l\boldsymbol{i} + m\boldsymbol{j} + n\boldsymbol{k})] \\ &\times [(x\boldsymbol{i} + y\boldsymbol{j} + z\boldsymbol{k}) \bullet (l'\boldsymbol{i} + m'\boldsymbol{j} + n'\boldsymbol{k})]\, \rho\, dv \end{aligned} \quad (4.16)$$

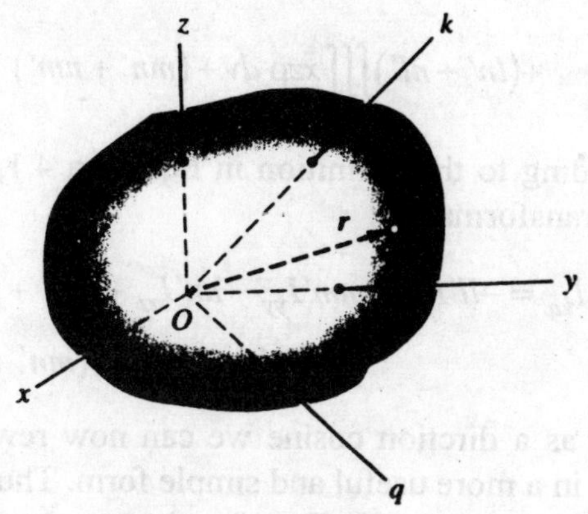

Figure 4.12 : *Find I_{kq}*

On carrying out the dot products in the integrand above, we obtain the following result :

$$I_{kq} = \iiint_V (xl + ym + zn)(xl' + ym' + zn')\, \rho\, dv$$

Therefore,

$$I_{kq} = \iiint_V \begin{pmatrix} x^2 ll' + y^2 mm' + z^2 nn' + xylm' + xzln' \\ + yxml' + yzmn' + zxnl' + zynm' \end{pmatrix} \rho\, dv \quad (4.17)$$

Noting $(ll' + mm' + nm')$ is zero from Equation 4.15 we may for convenience add the term.$(-x^2 - y^2 - z^2)$ (ll', mm', nn'), to the inergrand in the equation above. After, cancelation of some terms, we have

$$I_{kq} = \iiint_V \begin{pmatrix} -x^2 mm' - x^2 nn' - y^2 ll' - y^2 nn' - z^2 ll' - z^2 mm' \\ + yxlm' + xzln' + yxml' + yzmn' + zxnl' + zynm' \end{pmatrix} \rho\, dv$$

On, collecting terms and bringing the direction cosines outside the integrations, we obtain

$$I_{kq} = -ll' \iiint_V (y^2 + z^2)\, \rho\, dv - mm' \iiint_V (x^2 + z^2)\, \rho\, dv$$

$$-nn' \iiint_V (y^2 + x^2)\, \rho\, dv + (lm' + ml') \iiint_V xy\rho\, dv \quad (4.18)$$

$$+(ln' + nl')\iiint_V xz\rho\, dv + (mn' + nm')\iiint_V yz\rho\, dv$$

According to the definition in Equation 4.1, we can state the desired transformation:

$$I_{kq} = -ll'I_{xx} - mm'I_{yy} - nn'I_{zz} + (lm' + ml')I_{xy}$$
$$+ (ln' + nl')I_{xz} + (mn' + nm')I_{yz} \quad (4.19)$$

Using as a dirction cosine we can now rewrite the previous equation in a more useful and simple form. Thus, noting that $l' = a_{qx,}$ etc., we proceed as in Equation 4.14 to obtain

$$-I_{kq} = I_{xx}a_{kx}a_{qx} - I_{xy}a_{kx}a_{qy} - I_{xz}a_{kx}q_{qz}$$
$$- I_{yx}a_{ky}a_{qx} + I_{yy}a_{ky}a_{qy} - I_{yz}a_{ky}a_{qz} \quad (4.20)$$
$$- I_{zx}a_{kz}a_{qx} - I_{xy}a_{kz}a_{qy} + I_{zz}a_{kz}a_{qz}$$

Again the right side and easily be set forth by first putting down the matrix array of I_{ij} and then inseting the a's with easily determined subscripts while remembering to insert minus signs for off-diagonal terms.

Looking ahead : Tensors

In fig. 4.12 if we make the axis *Ok* an *X'* axis at o and using the direction cosines for this axis ($a_{x'x'}$, $a_{x'y'}$, $a_{x'z'}$) we can formulate $I_{x'x'}$ from equation 4.14 by a similar procedure , we can consider axis *Ok* to a *y'* axis at 0 and we can formulate $I_{z'z'}$ using for this axis dirction cosines *x'*, *y'*, *z'* at 0. from Eq. 4.14 The mass moments of inertia for reference for reference *x'*, *y'*, *z'* at o rotated arbitrarily relative to *xyz* can thus be obtained. We can evaluate *Ix'y'* at o using Equation 4.20 by considering the *Ok* and *Oq* axes to be *x'* and *y'* axes respectively, with *ax'x, ax'y,* and *ax'z* as direction cosines for the *x'* axis and *ay'x, ay'y* and *ay'z,* as direction cosines for the *y'*axis. This approach can similarly be followed to find $I_{x'z'}$ and $I_{y'z'}$. Thus, employing Equation 4.14 and 4.20 as , we can develop equations for computing the nine inertia quantities reference *x'y'z'* rotated arbitrarily relative to *xyz* at *O* in terms of the inertia quantities for reference *xyz* at zero. Thus, once the nine inertia quantities are known for one reference at some point, they can be determined for any reference at this point. By mean of certain transformation formed from Equation 4.14 and 4.20 the inertia terms transfors from one set of components for *xyz* at some point *O* to another set of components for *x'y'z'* at point *O*.

We now define a symmetric, *second-order tensor as a set of nine components.*

$$\begin{pmatrix} A_{xx} & A_{xy} & A_{xz} \\ A_{yx} & A_{yy} & A_{yz} \\ A_{zx} & A_{zy} & A_{zz} \end{pmatrix}$$

which transforms with a rotation of axes according to the following equations. For the diagonal terms,

$$\begin{aligned} A_{kk} = {} & A_{xx}a_{kx}^2 + A_{xy}a_{kx}a_{ky} + A_{xz}a_{kx}a_{kz} \\ & + A_{yx}a_{ky}a_{ky} + A_{yy}a_{ky}^2 + A_{yz}a_{ky}a_{kz} \\ & + A_{zx}a_{kz}a_{kx} + A_{zy}a_{kz}a_{ky} + A_{zz}a_{kz}^2 \end{aligned} \tag{4.21}$$

For the off-diagonal terms,

$$\begin{aligned} A_{kq} = {} & A_{xx}a_{kx}a_{qx} + A_{xy}a_{kx}a_{qy} + A_{xz}a_{kx}a_{kz} \\ & + A_{yx}a_{ky}a_{qx} + A_{yy}a_{ky}a_{qy} + A_{yz}a_{ky}a_{kz} \\ & + A_{zx}a_{kz}a_{qx} + A_{zy}a_{kz}a_{qy} + A_{zz}a_{kz}a_{qz} \end{aligned} \tag{4.22}$$

On compairing Equations 4.21 and 4.22, respectively, with Equation 4.19 and 4.20, we can conclude that the array of terms

$$I_{ij} = \begin{pmatrix} I_{xx} & -I_{xy} & -I_{xz} \\ -I_{yx} & I_{yy} & -I_{yz} \\ -I_{zx} & -I_{zy} & I_{zz} \end{pmatrix}$$

is a second-order tensor.

Because of the common transformtion law identifying certain quantities as tensors, there will be extremely important common characteristics for these quantities which set them apart from other quantities. We become involved with tensors as an entity in the engineering sciences, physics, and applied mathematics in order to learn these common characteristics in an efficent way and to understand them better.

We haved shown an infinitesimal rectangular parallelepiped extracted from a solid under load to explore this point futher. On three orthogonal faces we have shown nine forces intensities (i.e. forces per unit area) Normal stresses are those with repeated indice while those with different pairs of indices are called *shear stresses*. Knowing nine such stresses, you can readily find three stresses, one normal and two orthogonal shear stresses, on any interface at any orientation inside the rectangular parallelpiped.Figure 4.14 shows that this occurs in a thin plate loaded in the plane of symmetry. Thus stress is a *second-order tensor*.

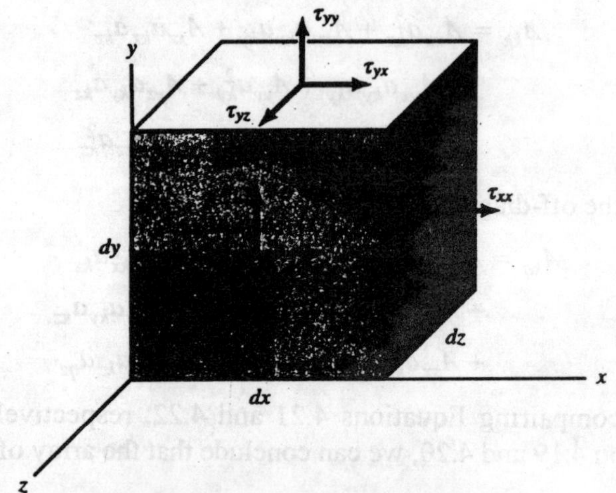

Figure 4.13 : ***Nine stresses on three orthogonal interfaces at a point.***

A two-dimensional simplification of τ_{ij} involving the quantities τ_{xx}, τ_{yy}, and $\tau_{xy}\left(=\tau_{yx}\right)$ as the only nonzero stresses is called *plane stress*. We have the same transformation equations given by Equations 4.21 and 4.22 to find such stresses on an interface knowing the stresses shown in figure 4.13. Plane stress is the direct analog of *second moments and products of area,*which is a two-dimensional simplification of the inertia tensor.

You will also learn that there are nine terms ε_{ij} that describe deformation at a point in solid mechanics. Thus consider the underformed an infintesimal rectangular parallelepiped shown inFigure 4.15. When there is a deformation there *normal strains* along the direction of the darkened edges which give the changes of length per unit original length of these edges. Furthermore, there

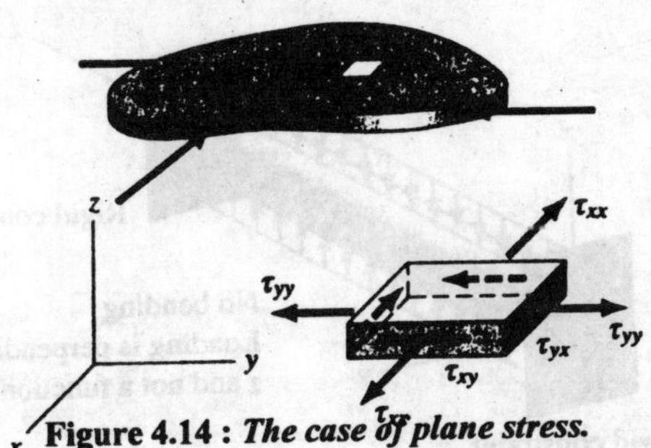

Figure 4.14 : ***The case of plane stress.***

are six *shear strains* $\varepsilon_{xy} = \varepsilon_{yx}, \varepsilon_{xz} = \varepsilon_{zx}, \varepsilon_{yz} = \varepsilon_{zy}$ that give the change in angle in radians from that of the right angles of three darkened edges which give the changes of length per unit original length of these edges, where there is a deformation. These other strains can be found by using transformation Equation 4.21 and 4.22 and so *strain is also a second-order tensor*.

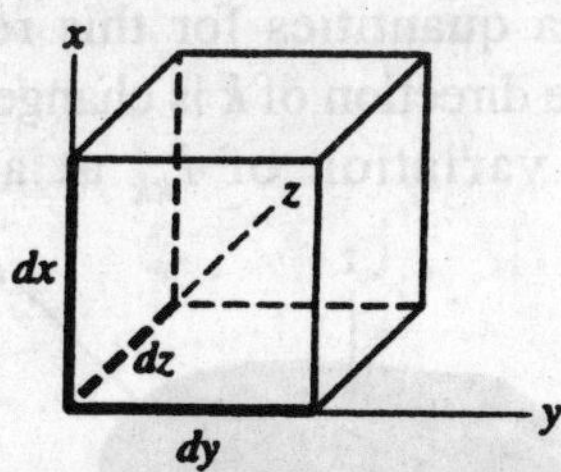

Figure 4.15 : ***An infinitesimal rectangular parallelepiped with three edges highlighted.***

The two-dimensional simplification of ε_{ij} involving the quantities ε_{xy}, ε_{yy}, and $\varepsilon_{xy}\left(=\varepsilon_{yx}\right)$ as the only nonzero strains is called *plane strain* and represents the strains in a prismatic body constrained at the ends with loading normal to the centerline in which the loading does not vary with z (see figure 4.16). Also, the prismatic body must not be subject to bending. Plane strain is an analogous mathematically to plain stress and second moments and products of area. All three are two-dimensional simplifications of *second-orer* symmetric tensors and have the *same transformation equations* as well as other mathematical properties. Finally, you will be introduced to quadruple tensor in electromagnetic theory and nuclear physics.

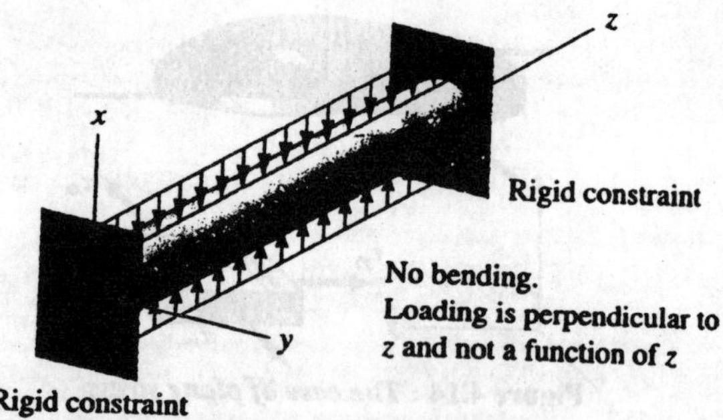

Figure 4.16 : *Examples of plane strain.*

The Inertia ellipsoid and principal moments of inertia

The moments of inertia of a body about an axis k in terms of the direction cosinesl of that axis measured form an orthogonal reference with an orighi O on the axis, and in terms of six independent inertria quantities for this reference is given by equation 4.14. As the direction of k is changed, we wish to explore the nature of the variation of I_{kk} at a point O in space.

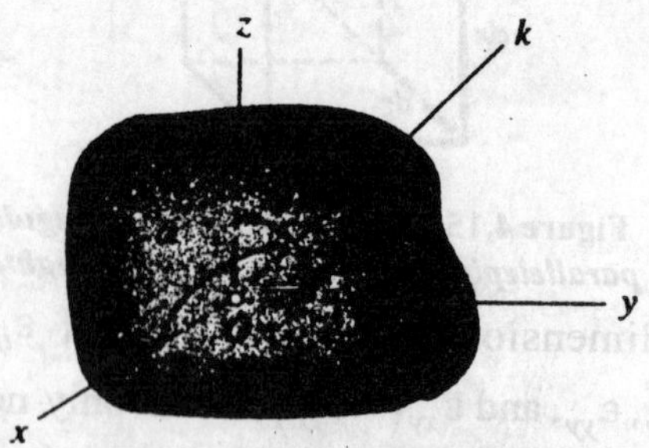

Figure 4.17 : *Physical diagram.*

(The k axis and the body are shown if Figure 4.17 which we shall call the physical diagram). We will employ a geometric representation of moment of inertia at a point that is developed in the following manner. We lay off as a distance the quantity OA given by the relation along the axis k.

$$OA = \frac{d}{\sqrt{I_{kk} / M}} \tag{4.24}$$

where d is and arbitrary constant that has a dimension of lenght that will render OA dimensionless, as it can be verified.The radius of gyration is $\sqrt{I_{kk} / M}$ and was presented earlier. To avoid confusion this operation is shown in another diagram, called the inertia diagram (Figure 4.18) where the new ξ, η, and ζ are *parallel the x, y, z,* axes of the physical diagram. We observe that some surface will be formed about the point O' by considering all possible direction so k, and this surface is related to the shape of the body through Equation 4.14. We can exprss the equation of this surface quite readily. The co-ordinate of point A are called ξ, η, and ζ. As $O'A$ is parallel to the line k and thus has the direction cosines a_{kx}, a_{ky}, and a_{kz} that are associated with this line, we can say that.

$$\begin{aligned} a_{kx} &= \frac{\xi}{O'A} = \frac{\xi}{d\sqrt{M / I_{kk}}} \\ a_{ky} &= \frac{\eta}{O'A} = \frac{\eta}{d\sqrt{M / I_{kk}}} \\ a_{kz} &= \frac{\zeta}{O'A} = \frac{\zeta}{d\sqrt{M / I_{kk}}} \end{aligned} \tag{4.25}$$

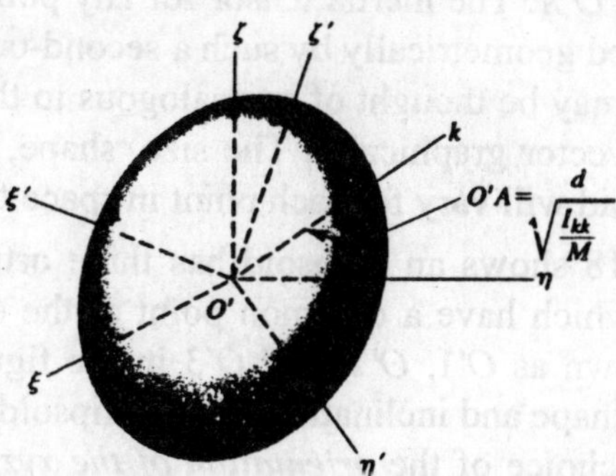

Figure 4.18 : ***Inertia diagram.***

Now, replace the direction cosines in Equation 4.13, using the relations above,

$$I_{kk} = \frac{\xi^2}{Md^2 / I_{kk}} I_{xx} + \frac{\eta^2}{Md^2 / I_{kk}} I_{yy} + \frac{\zeta^2}{Md^2 / I_{kk}} I_{zz}$$

$$+2\frac{\xi\eta}{Md^2 / I_{kk}}\left(-I_{xy}\right) + 2\frac{\xi\zeta}{Md^2 / I_{kk}}\left(-I_{xz}\right)$$

$$+2\frac{\eta\zeta}{Md^2 / I_{kk}}\left(-I_{yz}\right) \qquad (4.26)$$

In the preceeding equation we can see that I_{kk} cancels out leaving an equation involving the coordinates ξ, η, and ζ of the surface and the inertia terms of the body itself. Then rearranging the terms, we have

$$\frac{\xi^2}{Md^2 / I_{xx}} + \frac{\eta^2}{Md^2 / I_{yy}} + \frac{\zeta^2}{Md^2 / I_{zz}}$$

$$+\frac{2\xi\eta}{Md^2}\left(-I_{xy}\right) + \frac{2\xi\zeta}{Md^2}\left(-I_{zz}\right) + \frac{2\eta\zeta}{Md^2}\left(-I_{yz}\right) = 1 \qquad (4.27)$$

We know that the surface is that of an ellipsoid (see figure 4.18) through analytic geometry and is thus called the *ellipsoid of inertia.* The distance squared from O' to any point A on the ellipsoid is inversely propertional to the moment of inertia (see figure 4.24) about an axis in the body of O having the same directions as $O'A$. The inertia tensor for any point of a body can be represented geometrically by such a second-order surface, and this surface may be thought of as analogous to the arrow used to represent a vector graphically. The size, shape, and inclilnation of the ellipsoid will vary for each point in space for a given body.

Figure 4.18 shows an ellipsoid has three orthogonal axes of symmetry, which have a common point at the center, O', these axes are shown as $O'1$, $O'2$, and $O'3$ in the figure. We pointed out that the shape and inclination of the ellipsoid have nothing to do with the choice of the *orientation of the xyz,* and it depends on the mass distribution of the body about the origin of the *xyz* reference. (and hence the $\xi\eta\zeta$) reference at the point. We can therefore imagine that the *xyz* reference (and hence the $\xi\eta\zeta$ reference) can be chosen to have directions that coincide with the aforementioned symmetric axes $O'1$, $O'2$, $O'3$. If we call such references $x'y'z'$and $\xi'\eta'\zeta'$ respectively, we know from analytic geometry the Equation 4.27 becomes

$$\frac{(\xi')^2}{Md^2 / I_{x'x'}} + \frac{(\eta')^2}{Md^2 / I_{y'y'}} + \frac{(\zeta')^2}{Md^2 / I_{z'z'}} = \qquad (4.28)$$

where ξ', η', and ζ' are coordinates of the ellipsoidal surface relative to the new reference and $I_{x'x'}$, $I_{y'y'}$, $I_{z'z'}$ are mass moments of inertia of the body about the new axes. From the geometrical construction and the accompanying equation, we can now draw several important conlusions. One of the symmetrical axes of the dllipsoid above is the longest distance form the origin to the surface of the ellipsoid, and another axis is the smallest distance from the origin to the ellipsoidal surface. We must conclude that the minimum moment of inertia for the *O* must correspond to the axis having the maximum length, and the maximum moment of inertia must correspond to the axis having the minimum length by examining the definition in Equation 4.24. The third axis has an intermediate value that makes the sum of the moment of inertia tems equal to the sum of the moment of inertia terms for all orthogonal axea at point *O*, in accordance with Equation 4.2. In addition, equation 4.28 leads us to conclude that $I_{x'y'} = I_{y'z'} = I_{x'z'} = 0$. That is, the product of inertia of the mass about these axes must be zero. Clearly, these axes are the *principal axes* of inertia at the point *O*.

Since the preceding operations could be carried out at any point in space for the body, we can conclude that:

At each point there is a set of principal axes having extreme values of moments of inertia for that point and having zero products of inertia. The orientation of these axes will vary continuously from point to point throughout space for the given body.

All symmetric second-order tensor quantities have the

properties discussed above for the inertia tensor. We change the inertia tensor representaton form

$$\begin{pmatrix} I_{xx} & (-I_{xy}) & (-I_{xz}) \\ (-I_{yx}) & I_{yy} & (-I_{yz}) \\ (-I_{zx}) & (-I_{zy}) & I_{zz} \end{pmatrix} \text{ to } \begin{pmatrix} I_{x'x'} & 0 & 0 \\ 0 & I_{y'y'} & 0 \\ 0 & 0 & I_{z'z'} \end{pmatrix} \quad (4.29)$$

by transforming from the original reference to the principal reference .

In mathematical pr\arlance, we have "diagonalized" the tensor by the precedding operations.

5

Friction

Introduction

Whenever the surface of one body slides over the another body at rest each body exerts a force on the other. This force is called the frictional force or the force of friction. The frictional force is always opposite ot the relative motion of the body and never helps it. The property of the bodies by virtue of which a force is exerted by a stationary body on the moving body to resist the motion of the moving body is calle *friction.* Friction always acts parallel to the surface of contact and depends upon the nature of surface of contact.

Definitions

For defining the terms like co-efficient of friction (μ) and angle of friction (ϕ) consider a solid body placed on a horizontal plane surface as shown in Fugure 5.1.

Let W = Weight of body acting through C.G. downward,

R = Normal reaction of body acting through C.G. upward

P = Force acting on the body throught C.G. and paralle to the horizontal surface.

If P is small, the body will not move as the force of friction acting on the body in the dirction opposite to P will be more than P. But if the magnitude of P goes on increasing, a stage comes, when the solid body is on the point of motion. The force of friction acting on the body at this stage is called *limiting force of friction*. The limiting force of friction is denoted by F.

Resolving the forces on the body vertically and horizontally, we get

$$R = W$$

$$F = P.$$

If the magnitude of P is further increased the body will start moving. The force of friction, that comes into play when the two bodies in contact are in relative motion is called kinetic fricion.

Co-efficient of Friction (μ)

For any two surface in contact, it is the ratio of limiting force of friction and the normal reaction. It is denoted by μ. Thus,

$$\mu = \frac{\text{Limiting force of friction}}{\text{Normal reaction}} = \frac{F}{R}$$

$$F = \mu R \tag{5.1}$$

Angle of friction (ϕ)

It is the angle which the resultant of the force of limiting friction (F) and the normal reaction R makes with the normal reaction R. It is denoted by ϕ.

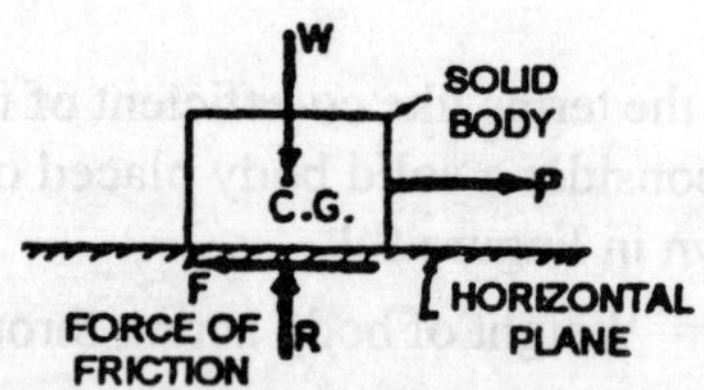

Figure 5.1 : *Solid body on horizontal surface.*

Let S = Resultant of the normal reaction (R) and limiting force of friction (F).

The angle of friction = ϕ = Anlge between S and R

From Figure 5.2, we have

$$\tan\phi = \frac{F}{R} = \frac{\mu R}{R} \qquad [\because F = \mu R \text{ from } (5.1)] \tag{5.1}$$

$$= \mu = \text{Co-efficient of friction} \tag{5.2}$$

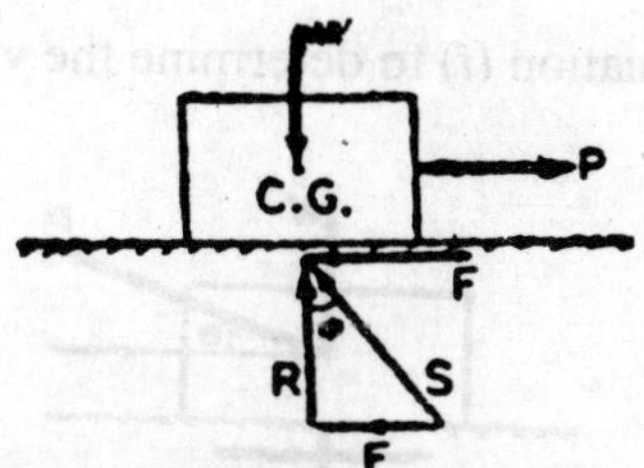

Figure 5.2 : ***Angle of friction.***

So, the tangent of the angle of friction is equal to the co-efficient of friction.

A block of weight W is placed on a rough horizontal plane surface as shown in Figure 5.3 and a force P ia applied at an angle θ with the horizontal such that the block just tends to move.

Let R = Normal reaction

μ = Co-efficient of friction

F = Force of friction

$= \mu R$

In the above case normal reaction R willnot be equal to the weight of the body. R is obtained by resolving the forces on the body in two components both horizontal and vertical. So force P is resolved in two components. $P \sin \theta$ in vertical direction $P \cos \theta$ in horizontal direction.

Resolving forces on the block horizontally, we get

$$F = P\cos\theta$$

or

$$\mu R = P\cos\theta \qquad (i)\ (\because F = \mu R)$$

Resolving forces on the block vertically, we get

$$R + P\sin\theta = W$$

$$R = W - P\sin\theta \qquad (ii)$$

From equation (*ii*) , it is clear that normal reaction is not equal to the weight of the block.

If in equation (*ii*), the values of W, P and θ are known, then value of normal reaction (R) can be obtained. This value of R can be substituted in equation (*i*) to determine the value of co-efficient of friction μ.

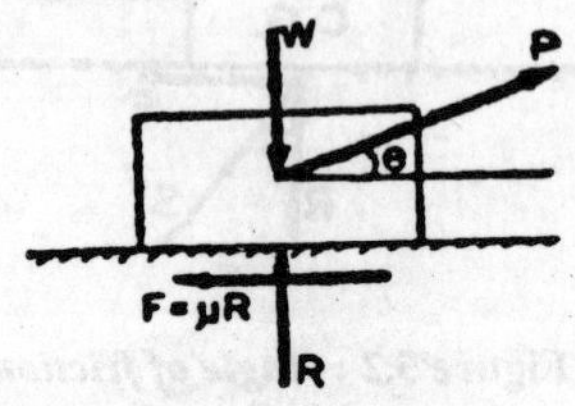

Figure 5.3

Cone of Friction

The right circular cone with vertex at the point of contact of the two surfaces, axis in the direction of normal reaction (*R*) and angle of friction (ϕ) equal to semi-vertical angle' is known as the cone of friction. Figure 5.4 shows the cone of friction in which,

O = Point of contact between two bodies

R = Normal reaction and also axis of the cone of friction

ϕ = Angle of frictoin.

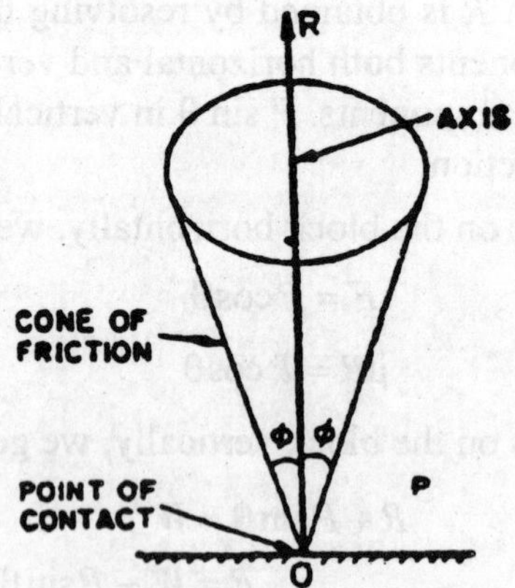

Figure 5.4 : ***Cone of friction.***

Types of friction

Depending upon the nature of contact between the two surfaces, friction is divided into following two types :-

1. Static friction, and
2. Dynamic friction.

When the two surfaces in contact are at rest then the force of friction which exactly balance the applied force is known as static friction but when the applied force exceeds the force of friction the body starts moving, the force experienced by the moving surface is called dynamic friction. If between the two surfaces, no lubrication (oil or grease) is used, the friction, that exists between two surface is called '*Solid Friction*' or '*Dry Friction*'.

Laws of Solid Friction

Solid friction is the force of friction that exist between the two surfaces which are not lubricated. The two surfaces may be at rest or one of the surface is moving and other surface is at rest. The following are the laws of solild friction :

1. The force of friction acts in the opposite direction in which surface is having tendency to move.
2. As long as the body is at rest, the force of friction is equal to the force applied to the surface.
3. When the surface is on the point of motion, the force of friction is maximum and this maximum frictional force is called the limiting friction force.
4. The limiting frictional force does not depend upon the shape and areas of the surfaces in contact.
5. The ratio between limiting friction and normal reaction is slightly less when the two surfaces are in motion.
6. The limiting frictional force bears a constant ratio to the normal reaction between two surfaces.
7. The force of friction is independent of the velocity of sliding.

The above laws of solid friction are also called laws of static and dynamic friction.

Angle of Repose

It is the angle that an inclined plane makes with the horizontal when a body placed on it remains in equilibrium by the assistance of friction only.

Consider a body of weight W, resting on a rough inclined plane as shown in Figure 5.5

Let R = Normal reaction acting at right angle to the inclined plane.

α = Inclination of the plane with the horizontal.

F = Frictional force acting upward along the plane.

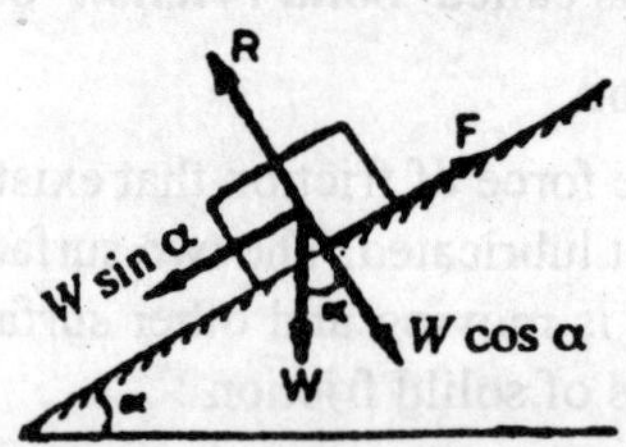

Figure 5.5

Increase the angle of inclination very slowly till the block just begins to slilde down the plane. This particular value of the angle of inclination is called the angle of repose.

Relsolving the forces along the plane, we get

$$W \sin\alpha = F \qquad (i)$$

Resolving the forces normal to the plane, we get

$$W \cos\alpha = R \qquad (ii)$$

Dividing equation (*i*) by equation (*ii*),

$$\frac{W \sin\alpha}{W \cos\alpha} = \frac{F}{R} \text{ or } \tan\alpha = \frac{F}{R} \qquad (iii)$$

But from equation (5.2),. we know

$$\tan\phi = \frac{F}{R} \qquad (iv)$$

where ϕ = angle of friction

Hence from equation (*iii*) and (*iv*), we have

$$\tan\alpha = \tan\phi$$

or $$\alpha = \phi$$

or Angle of repose = Angle of friction.

Equilibrium of a body lying on a rough inclined plane

If the angle of repose is less that the angle of friction the body tends to be in equilibrium without any external force.. If the body is to be moved upwards or downwards in the condition an external force is requird. But if the inclination of the plane or angle of repose is more than the angle of friction, the body will remain in equilibrium. The body will move downward and an upward and now an external force will be required to keep the body in equilibrium.

Such problems are solved by resolving the forces along the plane and perpendicular to the planes. The force of friction (F), which is always equal to μR is acting opposite to the direction of motion of the body.

Analysis of Ladder Friction

Figure 5.6 Shows a ladder AC resting of the ground and leaning against a wall.

Let R_A = Reaction of A

R_C = Reaction of C

F_A = Force of friction at A

$= \mu R_A$

F_C = Force of friction of C

$= \mu R_C$

The ladder has some weight due to its self weight or when some man stands on the ladder, the upper end A of the ladder tends to slip downwards, and hence the force of friction between the ladder and the vertical wall $F_A = \mu R_A$ will be acting upwards as shown in Figure 5.6. Similarly, the lower end C of the ladder will tend to move towards right and hence a force of friction between ladder and floor $F_C = \mu R_C$ will be action towards left.

For the system to be in equilibrium, the algebric sum of vertical and horizontal compoints of the forces must be zero. It is also necessary that the moments of all the forces about any point must be zero.

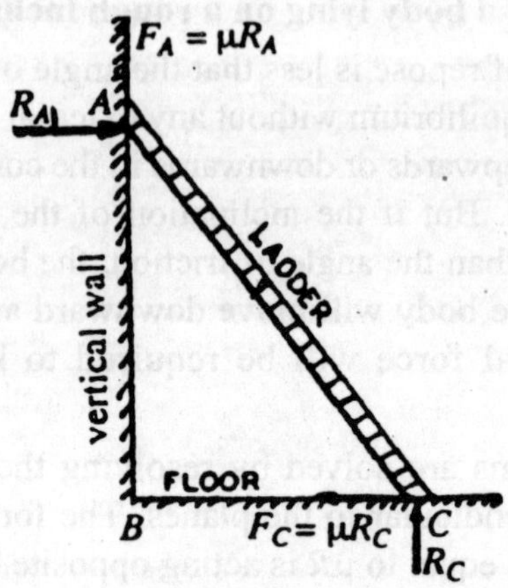

Figure 5.6

Note. *It the vertical wall is smooth, there will be no force of friction between the ladder and vertical wall.*

Screw-Jack

A device designed for lifting heavy wights or loads with the help of a small effort applied at its handle is known as Screw-jack. The followings are two types of screw-jack :

(*a*) Simple screw-jack, and

(*b*) Differential screw-jack.

(*a*) ***Simple screw-jack :*** Simple screw jack consists of a nut which forms the body of the jack, a screw with square threads and a handle fitted to the head of the the screw as shown in fig. 5.7.

The load W which is to be lifted is placed on the the head of the screw. At the end of the handel, fitted to the screw head, an effort P is applied in the horizontal direction to lift the load W. The screw-jack works on the principle on which an inclined plane works.

Let W = Weight placed on the screw head,

P = Effort applied at the end of the handle,

L = Length of handle,

p = Pitch of the screw,

d = Mean diameter of the screw,

α = Angle of the screw or helix angle,

ϕ = Angle of friction,

μ = Co-efficient of friction between screw and nut $= \tan\phi$

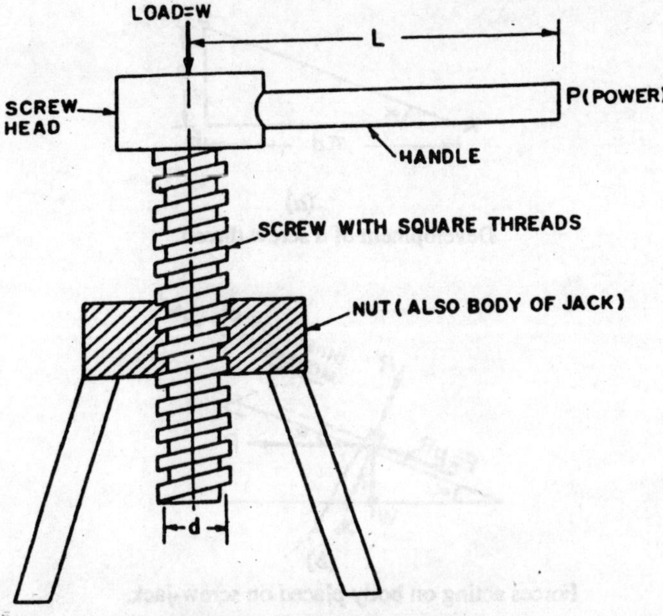

Figure 5.7 : *Simple screw-jack.*

When the handle is rotated through one complete turn, the screw is also rotated through one ture, and the load is lifted by a height p (pitch of screw).

The development of one complete turn of a screw thread is shown in Figure 5.8a. This is similar to the inclined plane. The distance AB will be equal to the circumference (πd) and distance BC will be equal to the pitch (p) of the screw. From the Figure 5.8 (a), we have

$$\tan\alpha = \frac{BC}{AC} = \frac{p}{\pi d} \qquad (5.3)$$

Let P' = Effort applied horizontally at the mean radius

of the screw-jack to lift the load W.

$r =$ Mean radius of the screw-jack $= \frac{d}{2}$.

Now this case becomes similar to that of lifting a load W up an inclined plane by a horizontal force P' as shown in Figure 5.8b

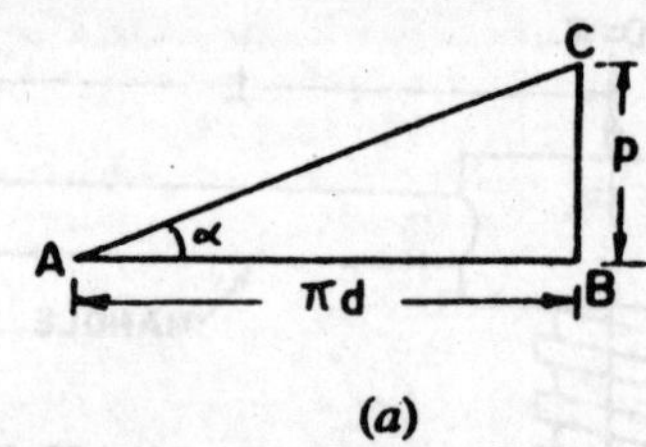

(a)

Development of a screw thread.

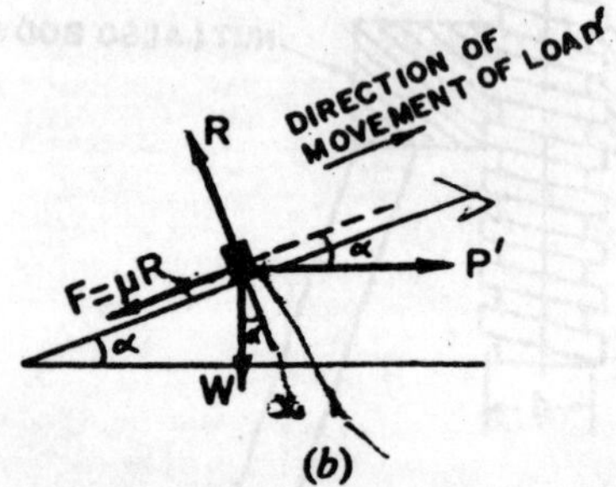

(b)

Forces acting on body placed on screw-jack.

Figure 5.8

Let $R =$ Normal reaction

$F =$ Force of friction μR

As the load W is lifted upwards, the force of friction will be acting downwards. All the forces acting on the body are shown in Figure 5.8 (b)

Resolving forces along the inclined plane,

$$F = W \sin\alpha = P' \cos\alpha$$

or $$\mu R + W \sin\alpha = P' \cos\alpha \quad (\because F = \mu R) \qquad (i)$$

Resolving forces normal to the inclined plane

$$R = W\cos\alpha + P'\sin\alpha$$

Substituting the value of R in equation (*i*), we get

$$\mu(W\cos\alpha + P'\sin\alpha) + W\sin\alpha = P'\cos\alpha$$

But $$\mu = \tan\phi = \frac{\sin\phi}{\cos\phi}$$

$$\frac{\sin\phi}{\cos\phi}(W\cos\alpha + P'\sin\alpha) + W\sin\alpha = P'\cos\alpha$$

or $$W\frac{\sin\phi\cos\alpha}{\cos\phi} + P'\frac{\sin\phi\sin\alpha}{\cos\phi} + W\sin\alpha = P'\cos\alpha$$

Multiplying by cos ϕ we get

$$W\sin\phi\cos\alpha + P'\sin\phi\sin\alpha + W\sin\alpha\cos\phi = P'\cos\alpha\cos\phi$$

or $$W[\sin\phi\cos\alpha + \sin\alpha\cos\phi] = P'[\cos\alpha\cos\phi - \sin\alpha\sin\phi]$$

or $$W\sin(\alpha+\phi) = P'\cos(\alpha+\phi)$$

$$P' = W\frac{\sin(\alpha+\phi)}{\cos(\alpha+\phi)} = W\tan(\alpha+\phi) \quad (5.4)$$

The effort applied at the mean radius of the screw-jack is P' but is case of screw-jack, effort is actually applied at the end of the handle as shown in Figure 5.7. P is the effort appllied at the end of the handle. Moment of P' about the axis of the screw

$= P' \times$ Distance of P' from the axis of the screw

$= P' \times$ Mean radius of the screw-jack

$= P' \times \frac{d}{2}$.

Moment of P about the axis of the screw

$= P \times$ Distance of P from axis

$= P \times L$.

Equating the two moments, we get

$$P' \times \frac{d}{2} = P \times L$$

$$P = P' \times \frac{d}{2L} = \frac{d}{2L} \times P' \quad (5.5)$$

Substituting the value of P' from equation (5.4) into equation (5.5), we get

$$P = \frac{d}{2L} \times W \tan(\alpha + \phi) \tag{5.6}$$

Equation (5.6) gives the relation between the effort required at the end of the handle and the load lifted.

Torque required to work the jack ***

Now

$$P = \frac{d}{2L} \times W \tan(\alpha + \phi)$$

$$= \frac{Wd}{2L} \frac{\tan\alpha + \tan\phi}{1 - \tan\alpha \tan\phi} \qquad \left(\because \tan(\alpha + \phi) = \frac{\tan\alpha + \tan\phi}{1 - \tan\alpha \tan\phi}\right)$$

$$= \frac{Wd}{2L} \frac{\frac{p}{\pi d} + \mu}{1 - \frac{p}{\pi d}\mu} \qquad \left(\because \tan\alpha = \frac{p}{\pi d}, \tan\phi = \mu\right)$$

$$= \frac{Wd}{2L}\left(\frac{p + \mu\pi\, d}{\pi d - p\mu}\right) \tag{5.7}$$

Equation (5.7) gives the value of P in terms of co-efficient of friction and pitch of the screw.

(*i*) *Effort Required at the end of the handle of the Screw-jack to lower the load W* : The screw-jack may also be used to lower the load. When the load is lowered by the screw-jack, the force by friction ($F = \mu R$) will act upwards. Figure 5.9 shows all the forces acting on the body.

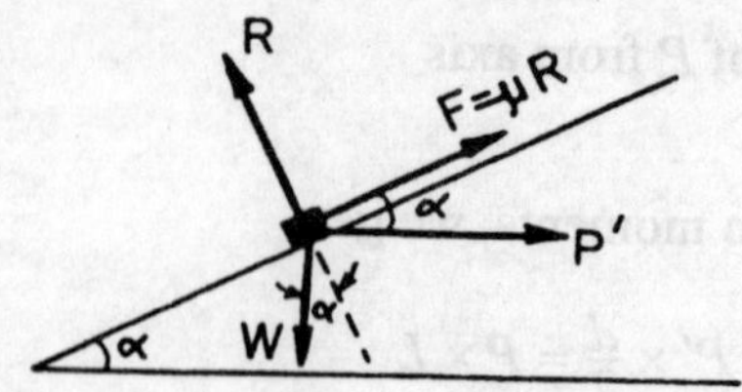

Figure 5.9 : *Body moving down.*

Resolving forces along the inclined plane

$$F = P'\cos\alpha = W\sin\alpha$$

or $$\mu R + P'\cos\alpha = W\sin\alpha \qquad (i)$$

Resolving forces normal to thc plane

$$R = W\cos\alpha + P'\sin\alpha$$

Substituting the value of R in equation (i), we get

$$\mu(W\cos\alpha + P'\sin\alpha) + P'\cos\alpha = W\sin\alpha$$

or $$\mu W\cos\alpha + \mu P'\sin\alpha + P'\cos\alpha = W\sin\alpha$$

or $$\mu P'\sin\alpha + P'\cos\alpha = W\sin\alpha - \mu W\cos\alpha$$

or $$P'[\mu\sin\alpha + \cos\alpha] = W[\sin\alpha - \mu\cos\alpha]$$

But $$\mu = \tan\phi = \frac{\sin\phi}{\cos\phi}$$

$\therefore$ $$P'\left[\frac{\sin\phi}{\cos\phi}\sin\alpha + \cos\alpha\right] = W\left[\sin\alpha - \frac{\sin\phi}{\cos\phi}\cos\alpha\right]$$

Multiplying by cos ϕ, we get

$$P'[\sin\phi\sin\alpha + \cos\alpha\cos\phi] = W[\sin\alpha\cos\phi - \sin\phi\cos\alpha]$$

or $$P'[\cos(\phi - \alpha)] = W[\sin(\phi - \alpha)]$$

$\therefore$ $$P' = W\frac{\sin(\phi - \alpha)}{\cos(\phi - \alpha)} = W\tan(\phi - \alpha) \qquad (5.8\text{ a})$$

If $$\alpha > \phi, \text{ then } P' = W\tan(\alpha - \phi) \qquad (5.8b)$$

But the effort applied at the mean radius of the screw-jack is P'. But in actual case, effort is applied at the handle of the jack. Let the effort at the handle is P as shown in Figure 5.8. Equating the moments of P and P' about the axis of the jack, we get

$$P \times L = P' \times \frac{d}{2}$$

$$P = \frac{d}{2L} \times P'$$

$$= \frac{d}{2L} \times W\tan(\phi - \alpha) \qquad [\because P' = W\tan(\phi - \alpha)] \qquad (5.8\text{ c})$$

Equation (5.8c) gives the relation between the effort required at the end of the handle to lower the load (W).

Expression for P in terms of co-efficient of fricton and pitch of the screw.

From equation (5.8c)

$$P = \frac{Wd}{2L}\tan(\phi - \alpha) = \frac{Wd}{2L}\frac{\tan\phi - \tan\alpha}{1 + \tan\phi\tan\alpha}$$

$$= \frac{Wd}{2L}\frac{\mu - \frac{p}{\pi d}}{1 + \mu\frac{p}{\pi d}} \qquad \left(\because \tan\phi = \mu, \tan\alpha = \frac{d}{\pi d}\right)$$

$$= \frac{Wd}{2L}\frac{\mu\pi d - p}{\pi d + \mu p} \tag{5.9}$$

(*ii*) Efficiency of a Screw-jack for rasing a Load W.

Let P = Actual effort required at the end of the handle of the screw jack to lift the load W.

W = Load lifted.

P_{ideal} = ideal effort required (or when there is no friction) at the end of the handle of the screw - jack to lift the load W.

d = Mean diameter of the screw.

L = Length of the handle

ϕ = Angle of friction

α = Angle of screw or helix angle.

μ = Co-efficient of friction = $\tan\phi$.

The actual effort (P) required at the end of the handle of the screw-jack to lift the load (W) is given be equation (5.6) as

$$P = \frac{d}{2L} W\tan(\alpha + \phi) \tag{i}$$

By making the friction zero, the ideal effort *(Pideal)* at the end of the handle of the screw jack to lilft the load (W) is obtained. Friction will be zero, when the co-efficient of friction $\mu = \tan\phi =$

0. This mean angle of friction ϕ is zero. If the value of $\mu = 0$ is substituted in equation (*i*), value of P becomes as P_{ideal}.

$$P_{ideal} = \frac{d}{2L} W \tan\alpha \qquad (ii)$$

The efficiency of a machine (here of screw-jack) in terms of ideal effort and actual effort is given by

$$\eta = \frac{\text{Ideal effort}}{\text{Actual effort}} = \frac{P_{ideal}}{P} \qquad (iii)$$

Substituting the values P_{ideal} from equation (*ii*) and the value of P from (*i*), in equation (*iii*), we have

$$\eta = \frac{\frac{d}{2L} W \tan\alpha}{\frac{d}{2L} W \tan(\alpha+\phi)} = \frac{\tan\alpha}{\tan(\alpha+\phi)} \qquad (5.10)$$

Equation (5.10) shows that the efficiency of a screw-jack in independent of the weight lifted or effort. The equation (5.10) can also be written as :

$$\eta = \frac{\frac{\sin\alpha}{\cos\alpha}}{\frac{\sin(\alpha+\phi)}{\cos(\alpha+\phi)}} = \frac{\sin\alpha}{\cos\alpha} \times \frac{\cos(\alpha+\phi)}{\sin(\alpha+\phi)}$$

Subtracting each side of the above equation from one,

$$-\eta = 1 - \frac{\sin\alpha \times \cos(\alpha+\phi)}{\cos\alpha \sin(\alpha+\phi)}$$

$$= \frac{\cos\alpha \sin(\alpha+\phi) - \sin\alpha \cos(\alpha+\phi)}{\cos\alpha \sin(\alpha+\phi)}$$

$$= \frac{\sin[(\alpha+\phi)-\alpha]}{\cos\alpha \sin(\alpha+\phi)} = \frac{\sin\phi}{\cos\alpha \sin(\alpha+\phi)}$$

$$= \frac{2\sin\phi}{2\cos\alpha \sin(\alpha+\phi)} \text{ (Multiplying and dividing by two)}$$

$$= \frac{2\sin\phi}{\sin[(\alpha+\phi)+\alpha] + \sin[(\alpha+\phi)-\alpha]}$$

$$= \frac{2\sin\phi}{\sin(2\alpha+\phi)+\sin\phi} \quad [\because 2\cos A\sin B = \sin(A+B)+\sin(A-B)]$$

Now η will be maximum if the term (1 – η) is minimum. But the term (1 – η) will be minimum if the denominator of R.H.S. of the above equation is maximum. Or in other words the value of [sin (2α + ϕ) + sin ϕ] is maximum. But ϕ (the angle of friciton) is constant and hence [sin (2α + ϕ) + sin ϕ] for a constant value of ϕ will be maximum if sin (2α + ϕ) is maximum. This is only possible when

$$2\alpha+\phi = 90° \quad \text{or} \quad 2\alpha = 90°-\phi$$

or

$$\alpha = 45°-\frac{\phi}{2}.$$

So, we can say, the condition for maximum efficiency is that $\alpha = 45°-\frac{\phi}{2}$. By substituting the above value of α in equation (5.10), maximum efficiency is obtained as :

$$\eta_{max} = \frac{\tan\left(45°-\frac{\phi}{2}\right)}{\tan\left(45°-\frac{\phi}{2}+\phi\right)} \quad \left(\because \alpha = 45°-\frac{\phi}{2}\right)$$

$$= \frac{\tan\left(45°-\frac{\phi}{2}\right)}{\tan\left(45°-\frac{\phi}{2}\right)} = \frac{\left(\dfrac{\tan 45°-\tan\frac{\phi}{2}}{1+\tan 45°\tan\frac{\phi}{2}}\right)}{\left(\dfrac{\tan 45°+\tan\frac{\phi}{2}}{1-\tan 45°\tan\frac{\phi}{2}}\right)}$$

$$= \frac{\left(\dfrac{1-\tan\frac{\phi}{2}}{1+\tan\frac{\phi}{2}}\right)}{\left(\dfrac{1+\tan\frac{\phi}{2}}{1-\tan\frac{\phi}{2}}\right)} \quad (\because \tan 45° = 1)$$

$$= \frac{\left(1-\tan\frac{\phi}{2}\right)}{\left(1+\tan\frac{\phi}{2}\right)} \times \frac{\left(1-\tan\frac{\phi}{2}\right)}{\left(1+\tan\frac{\phi}{2}\right)} = \frac{\left(1-\tan\frac{\phi}{2}\right)^2}{\left(1+\tan\frac{\phi}{2}\right)^2}$$

$$= \frac{\left[1-\frac{\sin\frac{\phi}{2}}{\cos\frac{\phi}{2}}\right]^2}{\left[1+\frac{\sin\frac{\phi}{2}}{\cos\frac{\phi}{2}}\right]^2} = \frac{\left[\cos\frac{\phi}{2}-\sin\frac{\phi}{2}\right]^2}{\left[\cos\frac{\phi}{2}+\sin\frac{\phi}{2}\right]^2}$$

$$= \frac{\cos^2\frac{\phi}{2}+\sin^2\frac{\phi}{2}-2\cos\frac{\phi}{2}\sin\frac{\phi}{2}}{\cos^2\frac{\phi}{2}+\sin^2\frac{\phi}{2}+2\cos\frac{\phi}{2}\sin\frac{\phi}{2}}$$

$$= \frac{1-2\cos\frac{\phi}{2}\sin\frac{\phi}{2}}{1+2\cos\frac{\phi}{2}\sin\frac{\phi}{2}} \quad \left(\because \cos^2\frac{\phi}{2}+\sin^2\frac{\phi}{2}=1\right)$$

$$= \frac{1-\sin\phi}{1+\sin\phi} \quad \left(\because \sin\phi = 2\cos\frac{\phi}{2}\sin\frac{\phi}{2}\right).$$

$$\therefore \qquad \eta_{max} = \frac{1-\sin\phi}{1+\sin\phi} \qquad (5.11)$$

(*b*) ***Differential Screw-jack*** : Figure 5.10 shows a differential screw-jack.

Differential screw consists of two parts *A* and *B*. Part *A* is threaded both on outside and inside : whereas the part *B* is threaded on the outside only. The outer or external threads of *A* gear with the thread of nut *C* which forms the body of differential screw jack. The external threads of screw *B* gears with internal threads of *A*. Thus part *A* behaves as screw for nut *C* and as nut for screw *B*.

The principle on which differential screw-jack works is: Action of one part of the machine is subtracted from the action of another part.

The screw *B* does not rotate, but moves verticall only, and carries the load. When the effort is applied at the lever, the screw *A* rises up and simultaneously the screw *B* goes down. Thus the net life of the load is algebraic sum of the motions of the screws *A* and screw *B*.

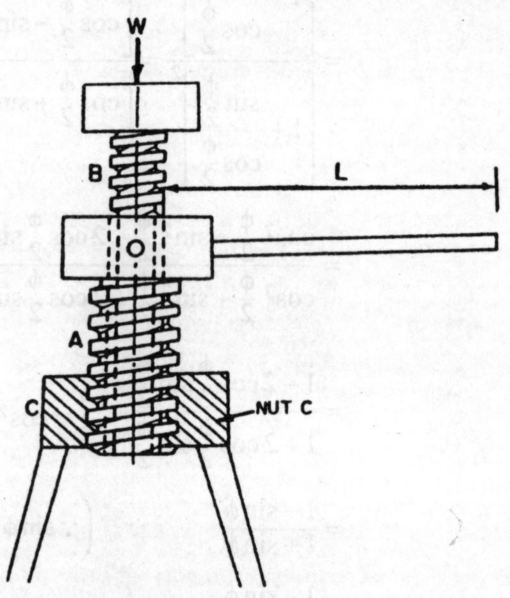

Figure 5.10

Let P_1 = Pitch of the screw *A*,

P_2 = Pitch of the screw *B*,

l = Length of the lever arm,

W = Load lifted, and

P = Effort applied to lift the load, at the end of the lever.

Consider one revolution of the lever are.

$\therefore$ Distance moved by the effort = $2\pi l$

Upward distance moved by $A = P_1$

Downward distance moved by $B = P_2$

Therefore the distance through which the load is lifted, $= P_1 - P_2$

$$\therefore \quad \text{Velocity ratio} = \frac{\text{Distance by the effort}}{\text{Distance moved by the load}} = \frac{2\pi l}{p_1 - p_2}$$

Belt Friction

A belt which is passing over a pulley is in contact with the surface of the pulley. If the suface of pulley is perfectly smooth, the tension throughout the belt will be constant. Also there will be no frictional resistance and so no driving torque will be developed.

The tension in the belt will not be constant if the surface of the pulley is rough. The tension will vary throughout the length of the belt which is in contact with pulley. Frictional resistance is responsible for this variation in tension. The frictional resistance depends on the co-efficient of friction (*i.e.*,value of μ) between the belt and pulley surface. It will be shown later that

$$\frac{T_1}{T_2} = e^{\mu\theta}$$

where T_1 = Tension in the belt on tight side,

T_2 = Tension in the belt on the slack side,

μ = Co-efficient of friction, and

θ = Angle of contact in radians

Ratio of belt tensions

Consier driver pulley *A* and driven pully *B* rotating in the clockwise direction. The given Figure shows only the driven pully *B*. Consider the driven pulley *B*.

The value of α is given by, $\sin\alpha = \dfrac{r_1 + r_2}{x}$ (5.12)

where r_1 = Radius of larger pulley,

r_2 = Radius of smaller pulley, and

x = Distance between the centres of the two pulleys.

Power Transmitted by Belt

Let T_1 = Tension in the tight side of the belt

T_2 = Tension in the slack side of the belt

v = Velocity of the belt in meter/s.

The difference between the two Tension (*i.e.,* $T_1 - T_2$) is the effective tension or force acting at the circumference of the driven pulley

∴ Effective driving force = $(T_1 - T_2)$

∴ Work done per second = Force × Velocity

$= (T_1 - T_2) \times v$ Nm

∴ Power transmitted $= \dfrac{(T_1 - T_2) \times v}{1000}$ kW (5.13)

or $P = (T_1 - T_2) \times v$ watts (5.14)

Equation (5.13) gives the power in *kW* whereas equation (5.14)gives the power in watts. In case of equation (5.14) the tensions T_1 and T_4 are taken in Newtons.

Torque exerted on the driving pulley $(T_1 - T_2) \times r_1$ (5.15)

and Torque exerted on the driven pulley $(T_1 - T_2) \times r_2$ (5.16)

Centrifugal Tension

Centrigugal tension is the tension caused by the centrifugal force in the running belt. Whenever a particle of mass *m* is ratated

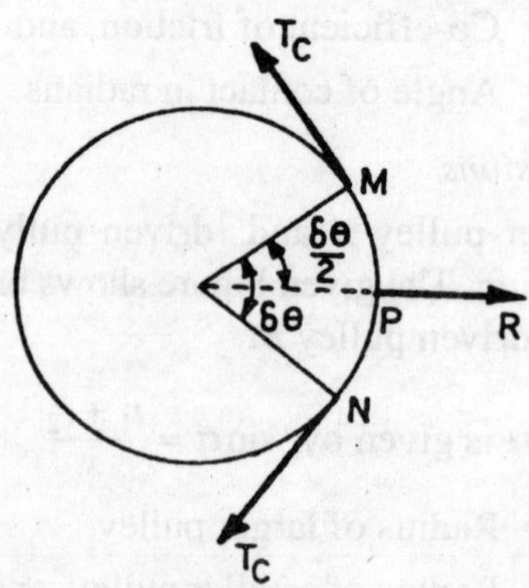

Figure 5.11

in a circular path of radius r at a uniform velocity v, a centirugal force is acting radially outward and its magnitude is equal to $\frac{mv^2}{r}$ where m is the mass of particle.

The centrifugal tension in the belt is calculatcd by considering the forces acting on an element on an elemental length of the belt (i.e., length MN) subtending an angle $\delta\theta$ at the centre oa shown in Figure 5.11

Let v = Velocity of belt in meter/s.

r = Radius of pulley over which belt runs

M = Mass of elemental length of belt

m = Mass of belt per meter length

T_c = Centrifugal tension acting at M and N tangentially

R = Centrifugal force acting radially outward.

The component of T_C acting radially inwards balance the centrifugal force R acting radially outward. Now elemental length of belt $MN = r \,.\, \delta\theta$

$\therefore$ Mass of belt MN = (Mass per meter length) × length of MN

or $$M = m \times r \times \delta\theta$$

$\therefore$ Centrifugal force, $$R = M \times \frac{v^2}{r} = m \times r \times \delta\theta \times \frac{v^2}{r}$$

Resolving the forces horizontally, we obtain,

$$T_c \sin\frac{\delta\theta}{2} + T_c \sin\frac{\delta\theta}{2} = R$$

or $$2T_c \sin\frac{\delta\theta}{2} = m \times r \times \delta\theta \times \frac{v^2}{r}$$

As the angle $\delta\theta$ is very small, so,

$$\sin\frac{\delta\theta}{2} = \frac{\delta\theta}{2}.$$

Then the above equation can be writter as:

$$2T_c \times \frac{\delta\theta}{2} = m \times r \times \delta\theta \times \frac{v^2}{r}$$

$$T_c = m \times v^2 \quad (5.17)$$

Note. (*i*) It is clear that the centrifugal tension is independent of and T_1 and T_2. From the above equation it depends upon the velocity of the belt. For lower belt speed (i.e., belt speed less than 10 m/s) the centrifugal tension is very small and may be neglected.

(*ii*) When the centrifugal tension is to be taken into consideration then total tensions on the right side and slack side of the belt is given as ;

For tight side $\quad T_1 + T_c \quad (5.18)$

For slack side $\quad T_2 + T_c \quad (5.19)$

(*iii*)Maximum tension (T_m) in the belt is equal to maximum safe stress in the belt multiplied by cross-sectional area of the belt.

$$T_m = f \times (b \times t) \tag{5.20}$$

where f = Maximum safe stress in the belt

b = Width of belt, and

t = Thickness of belt.

then $T_m = T_1 + T_c$ ... if centrifugal tension is to be considered

$= T_1$ ... if centrifugal tension is to be neglected

Maximum Power Transmitted by a Belt

Let T_1 = Tension on right side,

T_2 = Tension on slack side,

v = Linear velocity of belt,

Then the power transmitted is given be equation (5.14) as

$$P = (T_1 - T_2) \times v \qquad (i)$$

But from previou equation, We know that

$$\frac{T_1}{T_2} = e^{\mu \times \theta} \text{ or } T_2 = \frac{T_1}{e^{\mu \times \theta}}$$

Substituting the value of T_2 in equation (*i*), we get

$$P = \left(T_1 - \frac{T_1}{e^{\mu \times \theta}} \right) \times v$$

$$= T_1 \left(1 - \frac{1}{e^{\mu \times \theta}} \right) \times v \qquad (ii)$$

Let $\left(1-\frac{1}{e^{\mu\times\theta}}\right)=k$

Now, above equation becomes as

$$P = T_1 \times k \times v \text{ or } kT_1v \qquad (iii)$$

Let T_{max} = Maximum tension, and

T_c = Centrifugal tension which is equal of $m \times v^2$

Then $T_{max} = T_1 + T_c$

$\therefore \quad T_1 = T_{max} - T_c$

Substituting this value of T_1 in equation (*iii*), we get

$$P = k\,(T_{max} - T_c) \times v$$

$$= k\,(T_{max} - m \times v^2) \times v \qquad (\because T_c = m \times v^2)$$

$$= k\,[T_{max} \times v - m \times v^3) \qquad (iv)$$

The power transmitted will be maximum, if

$$\frac{d}{dv}(P) = 0$$

Now, differentiating equation (*iv*) w.r.t. *v* and equation to zero for maximum power, we get

$$\frac{d}{dv}(P) = k\,(T_{max} - 3\,m \times v^2) = 0$$

or

$$T_{max} - 3\,m \times v^2 = 0$$

$$T_{max} = 3\,m \times v^2 \qquad (v)$$

or

$$T_{max} = 3m\,v^2$$

or

$$v = \sqrt{\frac{T_{max}}{3m}} \qquad (5.21)$$

Equation (5.21) gives the velocity of the belt at which maximum power is tranmitted.

From equation (*v*), $T_{max} = 3\,T_c \qquad (\because m \times v^2 = T_c)$

Hence when the power transmitted is maximum, centrifugal tension would be $\frac{1}{3}$rd of the maximum tension.

We also know that $T_{max} = T_1 + T_c$

$$= T_1 + \frac{T_{max}}{3} \qquad \left(\because T_c = \frac{T_m}{3}\right)$$

$$\therefore \qquad T_1 = T_{max} - \frac{T_{max}}{3}$$

$$= \frac{2}{3} T_{max} \qquad (vi)$$

So, **Conditions** for the transmission of maximum power are :

(*i*) $T_c = \frac{1}{3} T_{max}$ and

(*ii*) $T_c = \frac{2}{3} T_{max}$.

Substituting the value of T_1 (from (*vi*))and value of belt velocity *v*(from equation 5.21) in equation (*ii*) we get the maximum power transmitted

$$\therefore \qquad \text{Maximum power} = \frac{2}{3} T_{max}\left(1 - \frac{1}{e^{\mu \times \theta}}\right) \times \sqrt{\frac{T_{max}}{3m}} \qquad (5.22)$$

6

Kinematics of Rigid Bodies: Relative Motion

Introduction

In previous chapter, we studied the kinematics of a particle. During virtually all of this study only a single reference was used. However, at the end of that chapter, we briefly introduced the use of two reference-for the case of *simple relative motion* involving two references *translating* with respect to each other.

In this chapter we well generalize the formulation of multireference analysis because of two reason :- Firstly by doing this we shall be able to analyse complicated motion in a more simple systematic way be using several references. Secondly, the motion of a particle is often known relative to a moving body (such as an airplane), to which we can fix a reference *xqz*, while the motion of the plane (and hence *xyz*) is known relative to an *internal reference XYZ*(such as the ground). Now **Newton's law** in the form $F = ma$, is valid only for an inertial reference. Hence, to use **Newton's law** for the particle reference directly. Accordingly, for practicle reasons we must become involved in multireference systems.

A reference is a rigid body, before we proceed further, we must firstly study the kinematics of a rigid body. In doing so, we will also set the platform for the study involving the dynamics of rigid bodies.

Translation and Rotation of Rigid Bodies

For the study of dynamic purposes, a rigid body is supposed

to be composed of particles having continuous distribution and fixed distance between each other. We shall profitably define once again two simple types of motion of a rigid body:

Translation. If a body moves in such a way that all the particles have same veocity at a time *t* relative to some reference., the body is said to be in *translation* relative to this reference at this time. The velocity of a translating body can vary with time and so can be represented as *V(t)*. Motion along a straight line does not neccessorily means tranlational motion.For example, the body shown in Figure 6.1 is in translation over the interval indicated because at each instant, each particle in the body has a common velocity. A straight line between two points of the body such as *ab* in Figure 6.1 always retains on orientation parallel to its *original* direction during the motion, is an important characteristics of trnaslational motion.

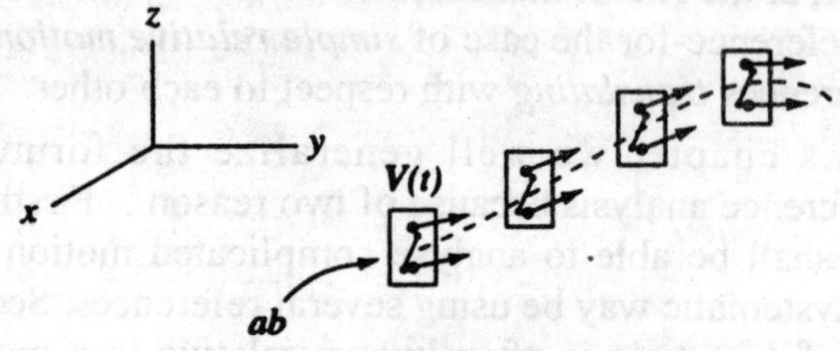

Figure 6.1 : *Translation of a body*

Rotation. If a rigid body moves so that along some straight line all the particles of the body, or a hypothetical extension of the

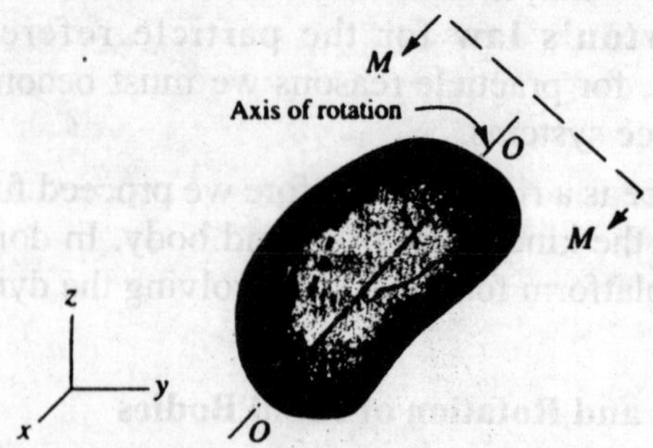

Figure 6.2 : *Rotation of a body*

body, hace zero velocity relative to some reference, the body is said to be in *rotation* relative to this reference. The *axis of rotation* is the line of stationary particles.

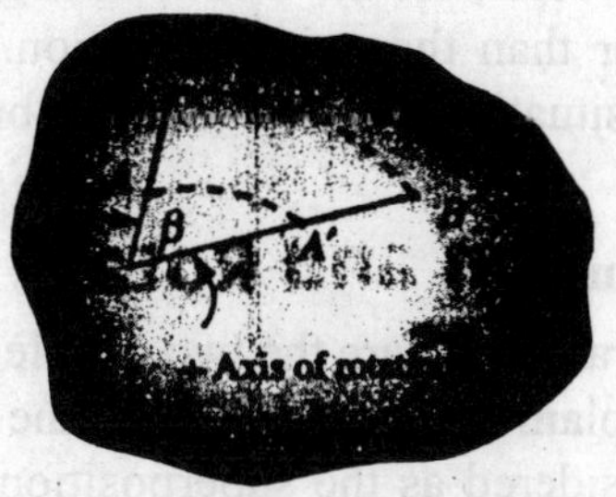

Figure 6.3 : *Measure of a partial rotation*

The amount of rotation in either a clockwise or a counterclock wise direction about the axis of rotation which brings the body back to the original position is defined as single revolution. We will now learn how to measure the rotation of a body. Partial revolutions can conveniently be measured by observing *any* line segment such as *AB* in the body (Figure 6.2)from a viewpoint *M-M* directed along the axis of rotation. This view of *AB* at the beginning of the partial rotation as seen along the axis of rotation , as well as the view *A'B'* at the end of the partial rotation is shown in Figure 6.3. The angle β that these lines form will be the same for the initial and final projections viewed along the axis of rotation of any line segment so examined in the partial rotation of the rigid body. Accordingly, the angle β so formed during a partial rotation is the measure of ratation.

In previous chapter, we studied, that finite rotations, although they have a magnitude and a direction along the axis of rotation, are not vectors. The superposition of rotations is not commutative, and therefore rotations can not be add according to the parallelogram law, which is a requirement of all vector quantities. However, if rotations become *infinitesimal*, they satisfy in the limit the commutative law of addition, so that infinitesimal rotations $d\beta$ are vector quantities. Therefore, the *angular velocity* is a vector quantity having a magnitude $d\beta/dt$ with an orientation parallel to the axis of ratation and a sense in accordance with the right-hand-screw rule. We shall employ ω to represent the angular velocity vector. Note that this difinition does not prescribe the line of action of this vector, for the line of action may be considered at positions other than the axis of rotation. The line of action depends on the situation at hand (as will be discussed in later sections).

Chasles' Theorem

Rotation and translation are the two simple motion of the body. Now we shall explain that at each instant the motion of any rigid body can be considered as the superposition of both rotational and translational motion.

For simplicity, consider a body moving in a plane. Positons of the body are shown tinted at times t and $(t + \Delta t)$in Figure 6.4. Now, let us select any point B of the body. The body is displaced without rotation form its position at time t to the position at time $(t + \Delta t)$ so that point B reaches its correct final position B'. ΔR_Bshows the displacement vector for this translation. We must now rotate the body an angle $\Delta\phi$ about an axis of rotation which is normal to the plane and which passes through point B' in order to reach the correct orientation for $(t + \Delta t)$.

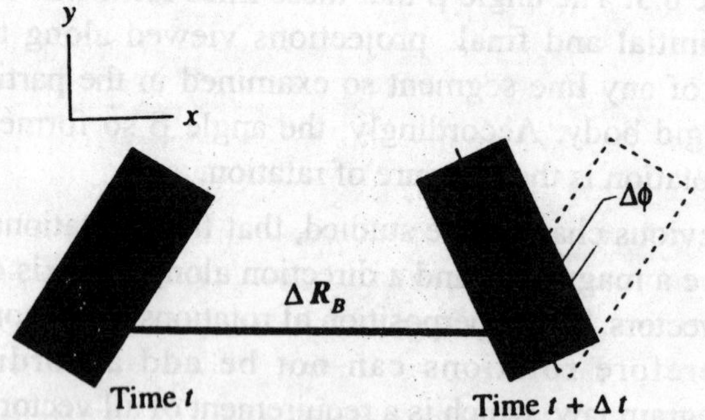

Figure 6.4 : ***Translation and rotation of a rigid body.***

What changes would occur had we chosen some other point C for such of procedure? Consider Figure 6.5, where we have included an alternative procedure by translating the body so that point C reached the correct final poition C'. In order to get the final orientation of the body we must rotate the body an amount

$\Delta\phi$ about an axis of rotation which is normal to the plane and which passes through C'. Thus two routes are indicated. We conclude from the diagram that the displacement $\Delta \boldsymbol{R}_C$ differs from $\Delta \boldsymbol{R}_B$ but there is no difference in the amout of rotation $\Delta\phi$. Thus, in general, ΔR *and the axis of rotation will depend on the point chosen, while the amount of rotation* $\Delta\phi$ *will be the same for all such points.*

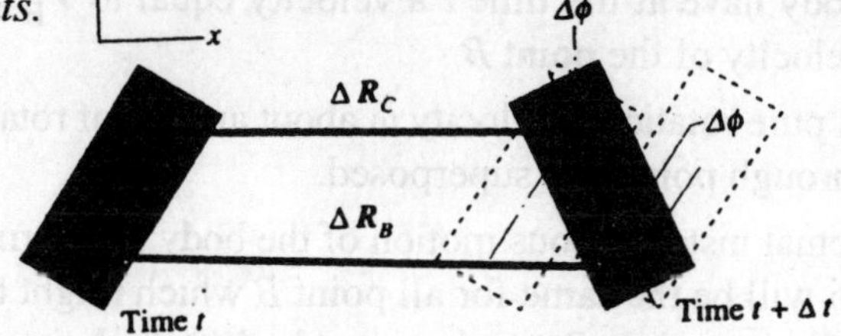

Figure 6.5 : *Translation and rotation of a rigid body using points* B *and* C.

Now, consider the ratios $\Delta \boldsymbol{R}/\Delta t$ and $\dfrac{\Delta\phi}{\Delta t}$. These quantities are regarded as an average translational velocity and an average rotational speed, respectively, of the body, which can be seperpose to get from the initial position to the final position in the time Δt. Thus $\Delta \boldsymbol{R}/\Delta t$ and $\Delta\phi/\Delta r$ represent an average measure of the motion during time interval Δt. We have instataneous translational and angular velocities which when superposed, give the instantaneous motion of the body if we go to the limit by letting $\Delta t \to 0$. The displacement vector of the chosen point B in the previous discussion represents the translation of the body during the time Δt. Furthermore, the chosen point B undergoes no other motion during Δt other than occuring during translation. Thus drawing conclusion we can say that, in the limit, the *translational velocity* used for the body corresponds to the *actual instantaneous* velocity of the chosen point B at time t. The angular velocity ω to be used in the movement of the body, is the same vector for all points B chosen, ω is the *instantaneous angular* velocity of the body.

We have thus far considered the movement of the body along

a plane surface. The same conclusions can be reached for the general motion of an arbitrary rigid body in space. The following statements for the description of the general motion of a rigid body relative to some reference at time t can be made. These statements comprise **Chasles' theorem.**

1. Select any point B in the body. Suppose all particles of the body have at the time t a velocity equal to V_B the actual velocity of the point B.
2. A pure rotational velocity ω about an axis of rotation going through point B is superposed.

The actual instantaneous motion of the body is determined with V_B and ω will be the same for all point B which might be chosen. When different points B are chosen only the translational velocity and the axis of rotation change. However, clearly understand that the *actual instantaneous axis of rotation* at time t is the one going through those points of the body having zeor velocity at time t.

Derivative of a Vector Fixed in a Moving Reference

In fig. 6.6 two references XYZ and xyz move arbitrarily relative to each other. Assume we are observing xyz from XYZ. We can apply **Chasles' theorem** to reference xyz since a reference is a rigid system. Thus, to fully describe the motion of xyz relative to XYZ, we choose O, and we superpose a translation velocity $\boldsymbol{R}$, equal to the velocity of O, onto a rotational velocity $\dot{\boldsymbol{R}}$, with an axis of rotation through O.

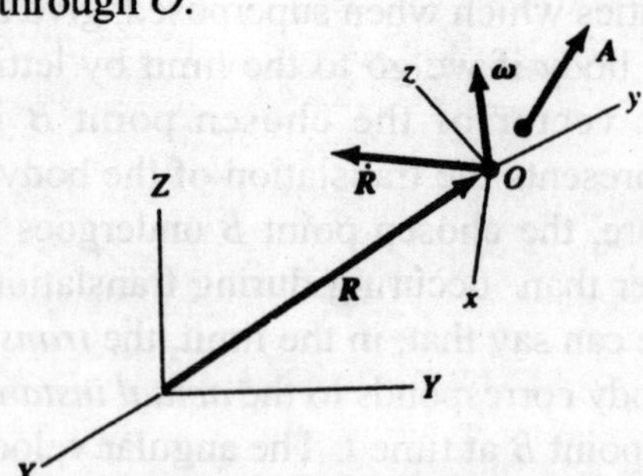

Figure. 6.6 : ***Vector A fixed in xyz moving relative of XYZ.***

As seen from reference xyz, assume that we have a vector $\boldsymbol{A}$ of *fixed length* and of fixed *orientation*. We say that a vector is "fixed" in reference *xyz*. Clearly, the time rate of change of *A* as seen from reference *xyz* must be zero. This statement can be expressed mathematically as :

$$\left(\frac{dA}{dt}\right)_{xyz} = 0$$

However, as seen from *XYZ*, the time rate of change *A* will not necessarily be zero. We make use of **Chasles' theorem** in the following manner, to evaluate $(dA/dt)_{xyz}$

1. Consider the *translational* motion **R**. This motion does not alter the direction of **A** as seen from *XYZ*. Thus vector **A** cannot change as a result of this motion as the magnitude of *A* is fixed.
2. We next consider *solely* a pure rotaiton about a stationary axis collinear with ω and passing throught point *O*.

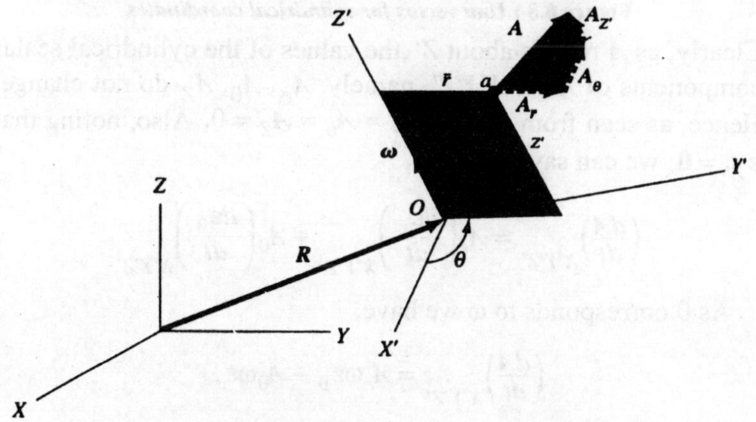

Figure 6.7 : *Cylindrical componentsw for vector A.*

We shall employ *O* a *stationary* reference *X'Y'Z'* positioned so that *Z'* coincides with the axis of rotation to best observe this rotation. This reference is shown in figure 6.7. Now at this instant the vector *A* is rotating about the '*Z*' axis. We have shown that cylindrical coordinates to the end of *A* (*i.e.* at point *a*); and have shown cylindrical components A_r, A_θ and $A_{z'}$, we have shown point *a* with unit vector ϵ_r, ϵ_θ, $\epsilon_{Z'}$, for cylindrical coordinates at this point in figure 6.8. We can accordingly express *A* as

$$A = A_r\varepsilon_r + A_\theta\varepsilon_\theta + A_{Z'}\varepsilon_{Z'}$$

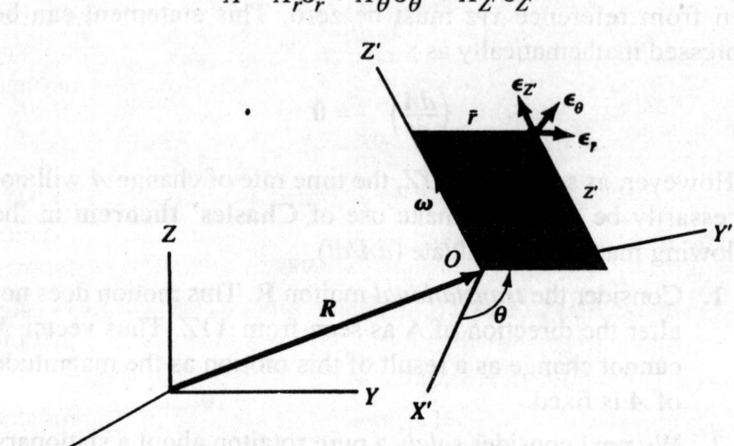

Figure 6.8 : ***Unit vectos for cylindrical coordinates.***

Clearly, as A rotates about Z' ,the values of the cylindrical scalar components of A for $X'Y'Z'$, namely, A_r, A_θ, $A_{Z'}$ do not change. Hence, as seen from $X'Y'Z'$, $\dot{A}_r = \dot{A}_\theta = \dot{A}_Z = 0$. Also, noting that $\dot{\varepsilon}_{Z'} = \mathbf{0}$, we can say that

$$\left(\frac{dA}{dt}\right)_{X'Y'Z'} = A_r\left(\frac{d\varepsilon_r}{dt}\right)_{X'Y'Z'} + A_0\left(\frac{d\varepsilon_\theta}{dt}\right)_{X'Y'Z'}$$

As θ corresponds to ω we have,

$$\left(\frac{dA}{dt}\right)_{X'Y'Z'} = A_r\omega\varepsilon_\theta - A_\theta\omega\varepsilon_r$$

But the right hand side is simply the cross product of ω and A as you can see by carrying out the cross product with cylindrical components. Thus,

$$\omega \times A = \omega\varepsilon_{Z'} \times \left(A_r\varepsilon_r + A_\theta\varepsilon_\theta + A_{Z'}\varepsilon_{Z'}\right)$$

$$= \omega A_r\varepsilon_\theta - \omega A_\theta\varepsilon_r$$

So, the conclusion is :

$$\left(\frac{dA}{dt}\right)_{X'Y'Z'} = \omega \times A$$

We would observe the same time derivative from the latter reference as from the former reference as $X'Y'Z'$ is stationary relative to XYZ. That is, $(d/dt)_{XYZ} = (d/dt)_{X'Y'Z'}$ and we can conclude that

$$\left(\frac{dA}{dt}\right)_{XYZ} = \omega \times A \tag{6.1}$$

The foregoing result gives the time rate of change of a vector A fixed in refercence xyz moving arbitrarily relative to reference XYZ. We see that $(dA/dt)_{XYZ}$ depends only on the vectores ω and A and not on their lines of action. Thus, we can conclude that the time rate of change of A fixed in xyz is not altered when :

1. The vector A is fixed at some other location in xyz provided the vector itself is not changed.
2. The actual axis of rotation of the xyz system is shifted to a new parallel position.

We can differntiate the terms in Equation 6.1 a second time. We thus obtain :

$$\left(\frac{d^2A}{dt^2}\right)_{XYZ} = \left(\frac{d\omega}{dt}\right)_{XYZ} \times A + \omega \times \left(\frac{dA}{dt}\right)_{XYZ} \tag{6.2}$$

Using Equation 6.1 to replace $(dA/dt)_{XYZ}$ and using $\dot{\omega}$ to replace $(d\omega / dt)_{XYZ}$ since the reference being used for this derivative is clear, we get

$$\left(\frac{d^2A}{dt^2}\right)_{XYZ} = \dot{\omega} \times A + \omega \times (\omega \times A) \tag{6.3}$$

We can compute higher-order derivatives by continuing the process. For this purpose only first equation i.e. $\left(\frac{dA}{dt}\right)_{XYZ} = \omega \times A$ must be remembered and all the subsequent higher order derivatives can be evaluated when needed.

So far, in this discussion we have considered a vector A, which is fixed in a reference xyz but this reference is a rigid can be replaced by a rigid body. The angular velocity ω used in Equation 6.1 is then the angular velocity of the rigid body in which A is fixed. We shall illustrate this condition in the following examples, which you are urged to study very carefully. For attaining a good working grasp of rigid body kinematics an understanding of these examples is necessary.

As an aid in carrying out computations involving the triple cross product, we wish to point out that the product

$$\omega_1 \boldsymbol{k} \times (\omega_1 \boldsymbol{k} \times C\boldsymbol{j}) = -\omega_1^2 C\boldsymbol{j}$$

That is, the product is minus the product of the scalars and has a direction corresponding to the last unit vector, ***j***.

In additon, consider a situation where the angular velocity of body *A* relative to body *B* is given by ω_1, while the angular velocity of body *B* relative to the ground is ω_2. What is the *total* angular velocity ω_T of body *A* relative to the ground. In such a case, we must remember that the angular velocity ω_1 of body *A* *relative* to body *B* is actually the *differnce* between the total angular velocity ω_2 of body *B* as seen form the ground. Thus

$$\omega_1 = \omega_T - \omega_2$$

Solving for ω_T, we ger

$$\omega_T = \omega_1 - \omega_2$$

It is clear from above that to get the total angular velocity ω_T, we simply add the various relative angular velocities just as we would with any pair of vectors.

Applications of the Fixed-Vector comcept

In previou section we considered the time derivative, as seen from a reference *XYZ*, of a vector ***A*** fixed in a rigid body or fixed in reference *xyz*. The result was a simple formula :

$$\dot{\boldsymbol{A}} = \underset{\sim}{\omega} \times \boldsymbol{A}$$

where ω is the angular velocity relative to *XYZ* of the body or the referenced in which ***A*** is fixed. In this part, we shall use the precceding formula for a *vector connecting two points a and b in a rigid body* (see figure 6.9).This vector if fixed in the body and is denoted by P_{ab} by us. The body has a velocity ***R*** relative to *XYZ* corresponding to some point *O* in the body plus an angular velocity ***R*** relative to *XYZ* with the axis of rotation going through *O* in accordance with the Chasels' theorem. We can then say on observing from *XYZ*:

$$\dot{\rho}_{ab} = \omega \times \rho_{ab} \tag{6.4}$$

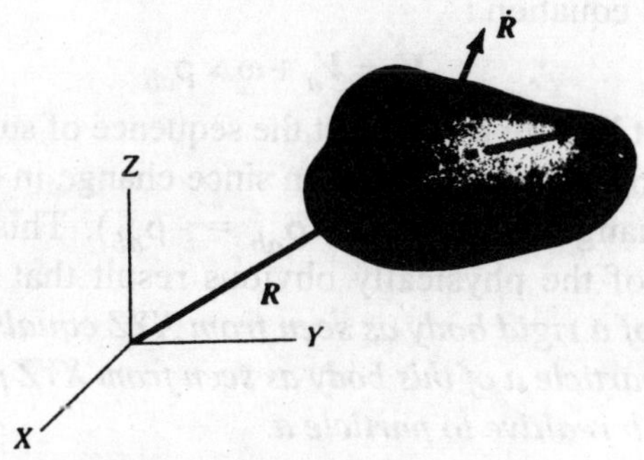

Figure 6.9 : ***Fixed in rigid body.***

Now, consider position vectore at a and b as shown in Figure 6.10. We can say:

$$\boldsymbol{r}_a + \rho_{ab} = \boldsymbol{r}_b$$

Taking the time derivative as seen from XYZ, we have

$$\left(\frac{d\boldsymbol{r}_a}{dt}\right)_{XYZ} + \left(\frac{d\rho_{ab}}{dt}\right)_{XYZ} = \left(\frac{d\boldsymbol{r}_b}{dt}\right)_{XYZ}$$

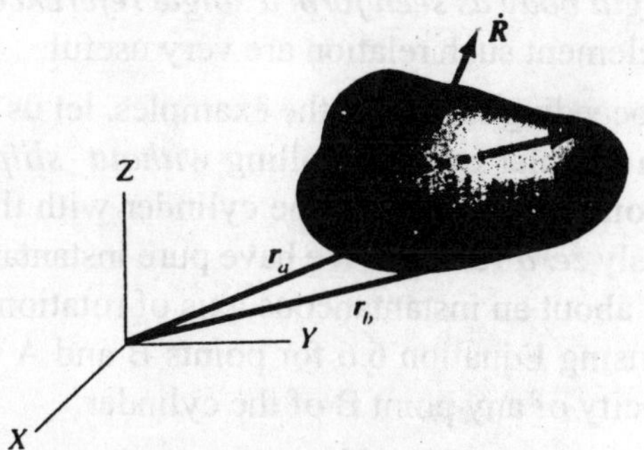

Figure 6.10 : ***Insert position vectors.***

This equation can be expressed as:

$$\left(\frac{d\rho_{ab}}{dt}\right)_{XYZ} = \boldsymbol{V}_b - \boldsymbol{V}_a \tag{6.5}$$

The difference between the velocity of point b and that at point a is $(d\rho_{ab} / dt)$ as noted above. We can say that $(d\rho_{ab} / dt)_{XYZ}$ is the velocity of point *b* *relative* to point *a*. Next, using Equation 6.4 to replace $(d\rho_{ab}/dt)_{XYZ}$, we have, on rearranging terms, a very useful equation :

$$V_b = V_a + \omega \times \rho_{ab} \tag{6.6}$$

We must be sure that we get the sequence of subscripts correct on ρ is using the above equation since change in ordering brings about a change in sign (i.e., $\rho_{ab} = -\rho_{ab}$). This equation is a statement of the physically obvious result that the *velocity of particel b of a rigid body as seen from XYZ equals the velocity of any other particle a of this body as seen from XYZ plus the velocity of particle b realtive to particle a.*

We can get a relation involving the accelearation vectors of two points on a rigid body by differentiating equation 6.6. again:

$$a_b = a_a + \left(\frac{d\omega}{dt}\right)_{XYZ} \times \rho_{ab} + \omega \times \left(\frac{d\rho_{ab}}{dt}\right)_{XYZ}$$

Hence, on using equation 6.4 in the last expression we have :

$$a_b = a_a + \dot{\omega} \times \rho_{ab} + \omega \times (\omega \times \rho_{ab}) \tag{6.7}$$

We have thus furmulated relations between *the motions of two points of a rigid body as seen form a single reference.* In the study of machine element such relation are very useful.

Before proceeding further to the examples, let us now consider the case of a circular cylinder rolling *without slilpping* (see fig 6.11). The point of contact*A* of the cylinder with the ground has instantaneously *zero velocity*. We have pure instantanious rotation at any time *t* about an instantaneous axis of rotation at the line of contact. By using Equation 6.6 for points B and A we can easily find the velocity of any point B of the cylinder.

$$V_B = V_A + \omega \times \rho_{ab}$$

Therefore,

$$V_B = 0 + \omega \times \rho_{AB} = \omega \times \rho_{AB}$$

From the above equation it is clear that for computing the velocity of any point on the cylinder we can thing of the cylinder *as hinged* at the point of contact. In particular for point *O*, the centerl of the cylinder, we get from above :

$$V_0 = -\omega\, Ri$$

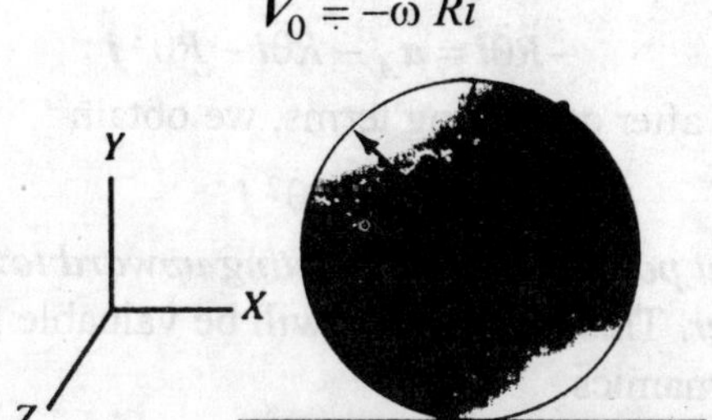

Figure 6.11 : *Cylinder rolling without slipping on a flat surface.*

The angular velocity has a magnitude of V_0/R if the velocity V_0 is known.

Another way of relating V and ω is to realize that the distance s and O moves must equal the length of circumference coming into contact with the ground. That is, measuring θ from the X axis to the Y axis:

$$s = -R\theta$$

On differentiating we obtain :

$$V_0 = -R\dot{\theta} = -R\omega$$

thus reproducing the previous result. Differentiating again, we obtain,

$$a_0 = -R\ddot{\theta} = -R\alpha \tag{6.8}$$

relating now the acceleration of O and the angular lacceleration α, the acceleration vector for O must be parallel to the ground. Again, we have a simple situation for computing a_0.

Next, for the point of contact A of the cylinder let us determine the accelearation vector. Thus, we can say for points A and O.

$$\boldsymbol{a}_O = \boldsymbol{a}_A + \dot{\omega} \times \rho_{AO} + \omega \times (\omega \times \rho_{AO})$$

Therefore,

$$-R\ddot{\theta}\boldsymbol{i} = \boldsymbol{a}_A + \ddot{\theta}\boldsymbol{k} \times R\boldsymbol{j} + \dot{\theta}\boldsymbol{k} \times \left(\dot{\theta}\boldsymbol{k} \times R\boldsymbol{j}\right) \tag{6.9}$$

Carrying out the products:

$$-R\ddot{\theta}\boldsymbol{i} = \boldsymbol{a}_A - R\ddot{\theta}\boldsymbol{i} - R\dot{\theta}^2\boldsymbol{j}$$

Therefore, after cancelling terms, we obtain :

$$\boldsymbol{a}_A = R\dot{\theta}^2\boldsymbol{j} \tag{6.10}$$

We see that point A is accelearating upward toward the center of the cylinder. This information will be valuable for the study of rigid body dynamics.

General Relationship between Time derivative of a Vector for different references

In previou section we considered the time derivatives of a vector ***A*** "fixed" in a reference *xyz* moving arbitrarily relative to *XYZ*. Our conclusions were :

$$\left(\frac{d\boldsymbol{A}}{dt}\right)_{xyz} = \boldsymbol{0}$$

$$\left(\frac{dA}{dt}\right)_{XYZ} = \omega \times A$$

We now aim to extend these considerations to include time derivatives of a vector ***A*** which is not necessarilly fixed in reference *xyz*. Primarily, our intention is to relate time derivative of such vector ***A*** as seen both form reference *xyz* and from*XYZ*, two reference moving arbitrarily relative to each other.

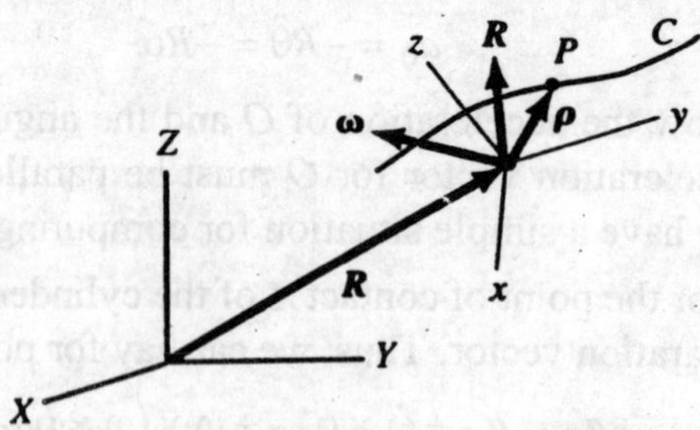

Figure 6.12 : *xyz moves relatives to XYZ.*

For this purpose, consider Figure 6.12 where a moving particle *P* with a position vector ρ in reference *xyz* is shown. According to the Chasles's theorem reference *xyz* means arbitrarily ralative to reference *XYZ* with translational velocity *R* and angular veloctiy ω. To relate the time derivative of any vector *A* as seen from any two reference we shall now draw a relation between $\left(\frac{d\rho}{dt}\right)_{xyz}$ and $\left(\frac{d\rho}{dt}\right)_{XYZ}$.

To obtain results effectively, we shall express the vector ρ in terms of components parallel to the *xyz* referecne:

$$\rho = x\boldsymbol{i} + y\boldsymbol{j} + z\boldsymbol{k} \tag{6.11}$$

where ***i, j,*** and ***k,*** are unit vectors for reference *xyz*. Differentiating this equation with respect to time for the *xyz* reference, we have:

$$\left(\frac{d\rho}{dt}\right)_{xyz} = \dot{x}\boldsymbol{i} + \dot{y}\boldsymbol{j} + \dot{z}\boldsymbol{k} \tag{6.12}$$

We must remember that ***i, j,*** and ***k*** of Equation 6.11 will be function of time, since these vectors will generally have some rotational motion relative to the *XYZ* reference if we next take the derivatives of ρ with respect to time for the *XYZ* reference. Thus, if dots are used for the time derivatives:

$$\left(\frac{d\rho}{dt}\right)_{XYZ} = (\dot{x}\boldsymbol{i} + \dot{y}\boldsymbol{j} + \dot{z}\boldsymbol{k}) + \left(x\dot{\boldsymbol{i}} + y\dot{\boldsymbol{j}} + z\dot{\boldsymbol{k}}\right) \tag{6.13}$$

The unit vector ***i*** is a vector *fixed* in reference *xyz*, and accordingly $\dot{\boldsymbol{i}}$ equals $\omega \times \boldsymbol{i}$. The same conclusions apply to ***j*** and ***k***. The last expression in parentheses can then be state as

$$\begin{aligned}\left(x\dot{\boldsymbol{i}} + y\dot{\boldsymbol{j}} + z\dot{\boldsymbol{k}}\right) &= x(\omega \times \boldsymbol{i}) + y(\omega \times \boldsymbol{j}) + z(\omega \times \boldsymbol{k}) \\ &= \omega \times (x\boldsymbol{i}) + \omega \times (y\boldsymbol{j}) + \omega \times (z\boldsymbol{k}) \qquad (6.14) \\ &= \omega \times (x\boldsymbol{i} + y\boldsymbol{j} + z\boldsymbol{k}) = \omega \times \rho\end{aligned}$$

In equation 6.13 we can replace $(\dot{x}\boldsymbol{i} + \dot{y}\boldsymbol{j} + \dot{z}\boldsymbol{k})$ by $(d\rho / dt)_{xyz}$, in accordance with Equation 6.12, and $\left(x\dot{\boldsymbol{i}} + y\dot{\boldsymbol{j}} + z\dot{\boldsymbol{k}}\right)$ by $\omega \times \rho$, in accordance with Equation 6.14. Hence,

$$\left(\frac{d\rho}{dt}\right)_{XYZ} = \left(\frac{d\rho}{dt}\right)_{xyz} + \omega \times \rho \tag{6.15}$$

We can generalize the preceding result for any vector A:

$$\left(\frac{dA}{dt}\right)_{XYZ} = \left(\frac{dA}{dt}\right)_{xyz} + \omega \times A \tag{6.16}$$

Where, ω *without subscripts will always be the angular velocity of the xyz reference relative to the XYZ reference.* Note that Equation 6.15 is a special case of Equation 6.16 since for A fixed in xyz, $(dA/dt)_{xyz} = \mathbf{0}$. We shall have much use for this relationship in succeeding sections.

Relationship between Velocities of a Particle for Different References

We shall now define the velocity of a particle again in the presence of several references.

The velocity of a particle relative to a reference is the derivative as seen from this reference of the position vector of the particle in the reference.

The velocities of the partice P relative to the XYZ and the xyz references in figure 6.12 are respectively,

$$\boldsymbol{V}_{XYZ} = \left(\frac{d\boldsymbol{r}}{dt}\right)_{XYZ}, \quad \boldsymbol{V}_{xyz} = \left(\frac{d\rho}{dt}\right)_{xyz} \tag{6.17}$$

$\boldsymbol{V}_{xyz}$ can be expressed in components parallel to the xyz reference at any time t while $\boldsymbol{V}_{xyz}$ may be expressed in components parallel to the XYZ reference at any time t as vector can always be decomposed into any set of orthogonal components.

Now, we shall relate these velocities by first noting that

$$\boldsymbol{r} = \boldsymbol{R} + \rho \tag{6.18}$$

Differentiating with respect to time for the XYZ reference, we get

$$\left(\frac{d\boldsymbol{r}}{dt}\right)_{XYZ} = \boldsymbol{V}_{XYZ} = \left(\frac{d\boldsymbol{R}}{dt}\right)_{XYZ} + \left(\frac{d\rho}{dt}\right)_{XYZ} \tag{6.19}$$

According to our definition, the velocity of the origin of the xyz reference relative to the XYZ reference is the term $(d\boldsymbol{R}/dt)_{xyz}$, and we denote this velocity as $\boldsymbol{R}$. The term $(d\rho/dt)_{XYZ}$ can be replaced, by use of Equation 6.15, in which $(d\rho/dt)_{xyz}$ is the velocity of the particle relative to the *xyz* refernce. Denoting $(d\rho/dt)_{xyz}$ simply $\boldsymbol{V}_{xyz}$ we find that the foregoing equation then becomes the desired relation:

$$\boldsymbol{V}_{XYZ} = \boldsymbol{V}_{xyz} + \dot{\boldsymbol{R}} + \omega \times \rho \tag{6.20}$$

ω represents the angular velocity of *xyz* relative to *XYZ*. This ω always goes into the above equation.

In Previous part of the chapter we dealt with the motion of two particles in a rigid body as seen form single reference. Now we shall consider the motion of single particle as seen from two references.

The multireference approach is very useful at time. For example, we could know the motion of a particle relative to some device, such as a rocket, to which we attach a reference *xyz*. Furthermore, form telemetering devices, we know the translational and rotatinal motion **(Chasles' theorem)** of the rocket (and hence *xyz*) relative to an inertial reference *XYZ*. It is important ot know the motion of the aforementioned particle relative directly to the inertial reference. The multireference approach clearly is invaluable for such problems.

The dot over a vector represents a time derivative as seen from *XYZ*.

Acceleration of a Particle for different References

The time derivative, as seen from the coordinate system, of the velocity relative of the coordinate system is the acceleration of a particle realtive to a co-ordinate system. Thus, observing Figure 6.13, we can say:

$$\begin{aligned} \boldsymbol{a}_{XYZ} &= \left(\frac{d}{dt}\boldsymbol{V}_{XYZ}\right)_{XYZ} = \left(\frac{d^2\boldsymbol{r}}{dt^2}\right)_{XYZ} \\ \boldsymbol{a}_{xyz} &= \left(\frac{d}{dt}\boldsymbol{V}_{xyz}\right)_{xyz} = \left(\frac{d^2\rho}{dt^2}\right)_{xyz} \end{aligned} \tag{6.21}$$

The equation first seem very typical to us, so now we will simplify it.

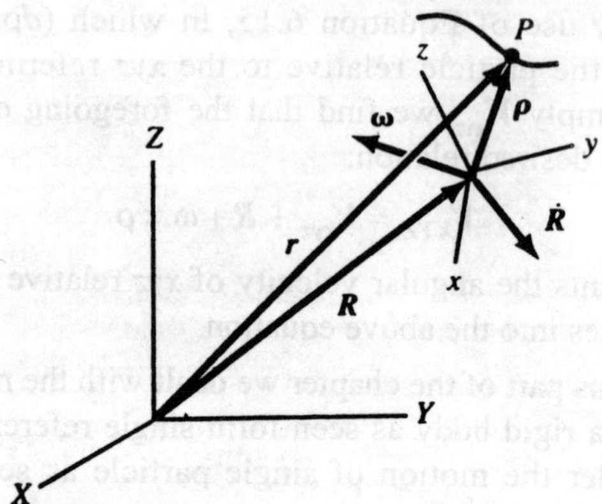

Figure 6.13 : *xyz moves arbitrarily to XYZ*

Now, let us relate the accelearation vectors of a particle for two references moving arbitrarily relative to each other. We do this by differentiating with respect to time the terms in Equation 6.20 for the *XYZ* reference. Thus,

$$\left(\frac{d\boldsymbol{V}_{XYZ}}{dt}\right)_{XYZ} \equiv \boldsymbol{a}_{XYZ} = \left(\frac{d\boldsymbol{V}_{xyz}}{dt}\right)_{XYZ} + \ddot{\boldsymbol{R}} + \left[\frac{d}{dt}(\omega \times \rho)\right]_{XYZ} \quad (6.22)$$

Now using the product rule carry out the derivative of the cross product:

$$\boldsymbol{a}_{XYZ} = \left(\frac{d\boldsymbol{V}_{xyz}}{dt}\right)_{XYZ} + \ddot{\boldsymbol{R}} + \omega \times \left(\frac{d\rho}{dt}\right)_{XYZ} + \left(\frac{d\omega}{dt}\right)_{XYZ} \times \rho \quad (6.23)$$

We can replace to introduce more physically meaningful terms

$$\left(\frac{d\boldsymbol{V}_{xyz}}{dt}\right)_{XYZ} \text{ and } \left(\frac{d\rho}{dt}\right)_{XYZ}$$

Now, using Equation 6.16 in the following way:

$$\left(\frac{d\boldsymbol{V}_{xyz}}{dt}\right)_{XYZ} = \left(\frac{d\boldsymbol{V}_{xyz}}{dt}\right)_{xyz} + \omega \times \boldsymbol{V}_{xyz}$$

$$\left(\frac{d\rho}{dt}\right)_{XYZ} = \left(\frac{d\rho}{dt}\right)_{xyz} + \omega \times \rho$$

Substituting into Equation 6.23, we get

$$a_{XYZ} = \left(\frac{dV_{xyz}}{dt}\right)_{xyz} + \omega \times V_{xyz} + \ddot{R} + \omega \times \left(\frac{d\rho}{dt}\right)_{xyz}$$

$$+ \omega \times (\omega \times \rho) + \left(\frac{d\omega}{dt}\right)_{XYZ} \times \rho$$

Thus, we note that $(dV_{xyz}/dt)_{xyz}$ is a_{xyz}, that $(d\rho/dt)_{xyz}$ is V_{xyz} and that $(d\omega/dt)_{xyz}$ is $\dot{\omega}$ .Hence, rearranging terms, we have

$$a_{XYZ} = a_{xyz} + \ddot{R} + 2\omega \times V_{xyz} + \dot{\omega} \times \rho + \omega \times (\omega \times \rho) \quad (6.24)$$

where ω and $\dot{\omega}$ are the angular velocity and acceleration, respectively, of the *xyz* reference relative to the *XYZ* reference. The vector $2(\omega \times V_{xyz})$ is called the *Coriolis acceleration vector.*

Equation 6.24 is harrible when first seen but it is helpful in solving those problems which would otherwise be tremendously difficult.

A new lool at Newton's law

The Newton's law in its proper form can be expressed as :-

$$F = ma_{xyz} \quad (6.25)$$

where a is the acceleration and is measured relative to an inertial reference. There are times when we know the motion of a particle and it make sense only relative to a non inertial reference. Such a case arises in an airplanr or rocket, where machine elements must move is a certain way relative to the vehicle in order to function properly. Therefore, the motion of the machine element relative to the vehicle is known. We can not use Equation 6.25 with the acceleration of the machine element measured relative to the vehicle if the vehicle is undergoing a severe maneuver relative to inertial space. This is so since the vehicle is not at that instant an inertial reference, and to disregard this fact will lead to wrong results. In such a problems, the motion of the vehicle may be known relative to inertial space, and we can employ to good advantage the multireference analysis of the previous section. We can then use Newton's law in the following ways attaching the reference *xyz* to the vehicle and *XYZ* to inertial space :

$$\boldsymbol{F} = m\left[\boldsymbol{a}_{xyz} + \ddot{\boldsymbol{R}} + 2\omega \times \boldsymbol{V}_{xyz} + \dot{\omega} \times \rho + \omega \times (\omega \times \rho)\right] \quad (6.26)$$

Clearly, the bracketed expression is the required quantity $\boldsymbol{a}_{XYZ}$ needed for Newton's law. It is the usual practice to write Equation 6.26 in the following form:

$$\boldsymbol{F} = m\left[\boldsymbol{a}_{xyz} + \ddot{\boldsymbol{R}} + 2\omega \times \boldsymbol{V}_{xyz} + \dot{\omega} \times \rho + \omega \times (\omega \times \rho)\right] = m\boldsymbol{a}_{xyz} \quad (6.27)$$

This equation is may considered as Newton's law written for a *noninertial* reference *xyz*. The terms $-m\ddot{\boldsymbol{R}}, -m(2\omega \times \boldsymbol{V}_{xyz})$, and so on, are then considered as forces and are termed *inertial forces.* Thus. we can take the viewpoint that for a noninertial references *xyz*, we can still say force ***F*** equals mass times acceleration $\boldsymbol{a}_{xyz}$ provided that we include with the applied force ***F***, all the inertial forces. Indeed, we shall adopt this viewpoint in this text. The inertial force $-2m\omega \times \boldsymbol{V}_{xyz}$ is the very interesting *Coriolis force.*

The inertial forces result in perplexing action that are sometimes just oppostie to ous intution. Most of us during our lives have been involved in actions where the reference used (knowingly or not) has been with sufficient accuracy an inertial reference, usually the earth's surface. We have, accordingly, become unconditioned to associating an acceleration proportional to, and in the same direction as, the applied force. But there are certain occasion when we find our motion reating to a highly non inertial reference. For example, fighter pilots and stunt pilots carry out actions in a cockpit of a plane while the plane in undergoing severe maneuvers. If flyers use the cockpit interior as a reference for their action than the result is quiet unexpected. Thus, to move their hands from one position to another relative to the cockpit sometimes requires an exertion that is not the one anticipated, causing considerable confusion.

The coriolis force

Previously defined Cariolis force is of great interest because it realtes to certain terrestrial actions. The earth's surface with sufficient accuracy serves as an inertial reference for many of our problems that we encounter. However, where the time interval of interest is large (such as in the flight of rockets, or the flows of rivers, or the movement of winds and ocean currents), we must consider such a reference as noninertial in certain instances and accordingly, when using **Newton's law**, we must include some or all of the inertial forces given in Equation 6.27. For such problems we generally use an inertial reference that has an origin at the center of the earth (see Figure 6.14) with the Z axis collinear with the N-S axis of the earth and moving such that the earth rotates one revolution per 24 hr relative to the reference. Thus, reference approaches a translatory motion about the sun. It is an inertial reference, to a high degree of accuracy.

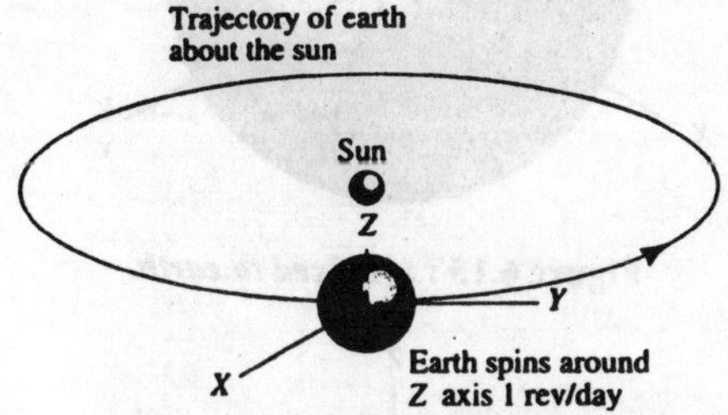

Figure 6.14 : ***Proposed inertial reference.***

We consider particles that are stationary relative to the earth. We choose a reference *xyz* fixed to the earth at the equator as shown in Figure 6.15. The angular velocity of *xyz* fixed anywhere on the earth's surface can readily be evaluated as follows :

$$\omega = \frac{2\pi}{(24)\,(3{,}600)}\boldsymbol{k} = 7.27 \times 10^{-5}\boldsymbol{k} \text{ rad / sec}$$

Newton's law, in the form of Equation 6.27, for a "stationary" particle positioned at the origing of *xyz* simplifies to

$$\boldsymbol{F} - m\ddot{\boldsymbol{R}} = \boldsymbol{0} \tag{6.28}$$

since ρ, $\boldsymbol{V}_{xyz}$ and $\boldsymbol{a}_{xyz}$ are zero vectors. Let us next evaluate the inertial force, $-m\ddot{\boldsymbol{R}}$ for the particle, using $R = 3{,}960$ mi:

$$-m\ddot{\boldsymbol{R}} = -m(-|\boldsymbol{R}|\,\omega^2\,\boldsymbol{j}) = m(3{,}960)\,(5{,}280)\,(7.27 \times 10^{-5})^2\,\boldsymbol{j}$$
$$= m\,(.1105)\,\boldsymbol{j} \text{ lb}$$

This is the centrifugal force. From the figure 6.16 we can see clearly that the direction of this force is collinear with the gravitational force acting on a particle but in a opposite sense. Further the centrifugal force has a magnitude that is (.1105 m/ 32.2 m) × 10 = .34 % of the gravitational force at the indicateсd location. Thus, in the usual engineering problems, such effects are neglected.

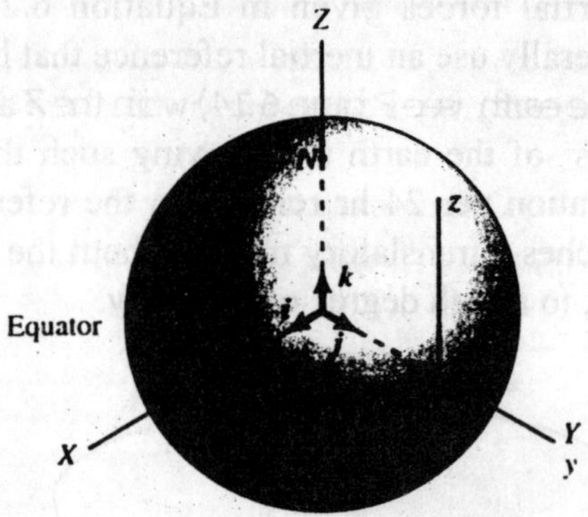

Figure 6.15 : *xyz fixed to earth.*

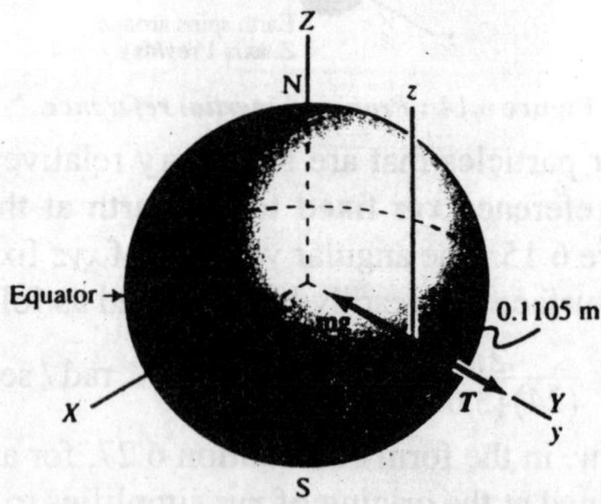

Figure 6.16 : *Plumb bob at equator.*

Now, let as assume that the particle is restrained by a flexible card form resting on the surface of the earth. In accordance with Equation 6.28, the external force ***F*** (which includes gravitational attraaction and the force from the cord) and the centrifugal force $-m\ddot{\boldsymbol{R}}$ add up to zero, and so these forces are in equilibrium. They are shown in Figure 6.16 in which *T* represents the contribution of the cord. Clearly, a force *T* radially out from the center of the earth will restrain the particle, and so the direction of the flexible cord will point toward the center of the earth. On the contrary this will not be true at a nonequatorial location. The gravity force points towards the center of the earth(see figure 6.17), but the centrifugal force-now having the value $m\left[\boldsymbol{R}(\sin\theta)\,\omega^2\right]$— points radially out from the Z axis, and thus *T*, the restraining force, must be inclined somewhat from a direction toward the center of the earth. Therefore, we see that *plumb bob* does not point direcly toward the center of the earth except at the equator of at the poles (where the contrifugal force is zero). This deviation is very small and is neglible for must but not all engineering work.

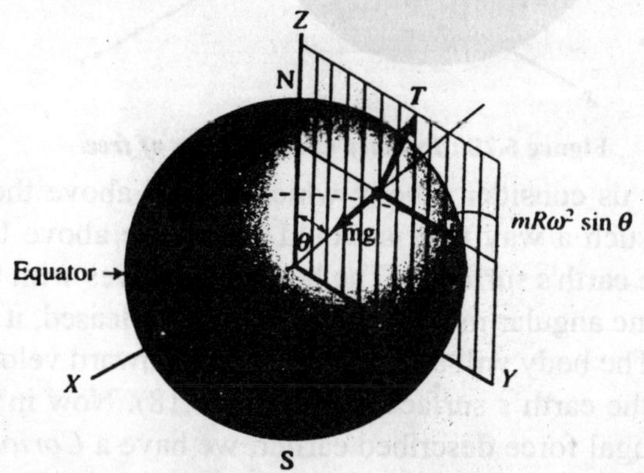

Figure 6.17 : *Plumb bob does not point of center of earth.*

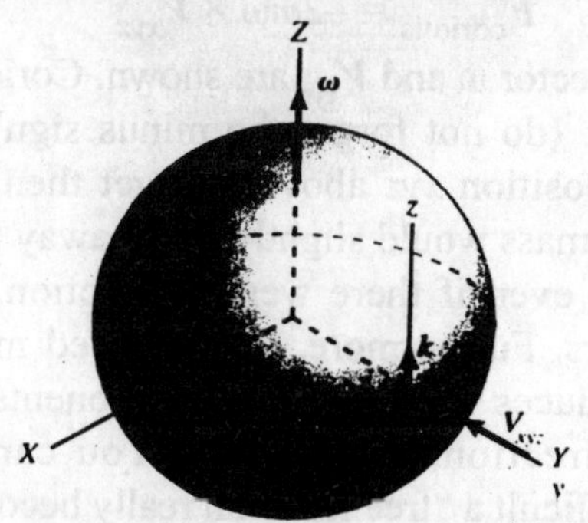

Figure 6.18 : *Free fall at the equator.*

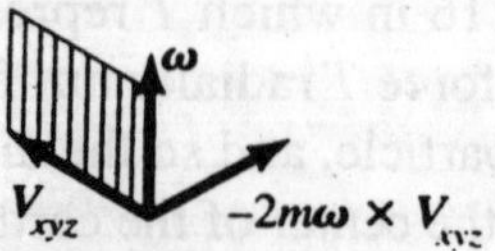

Figure 6.19 : *Direction of Coriolis force.*

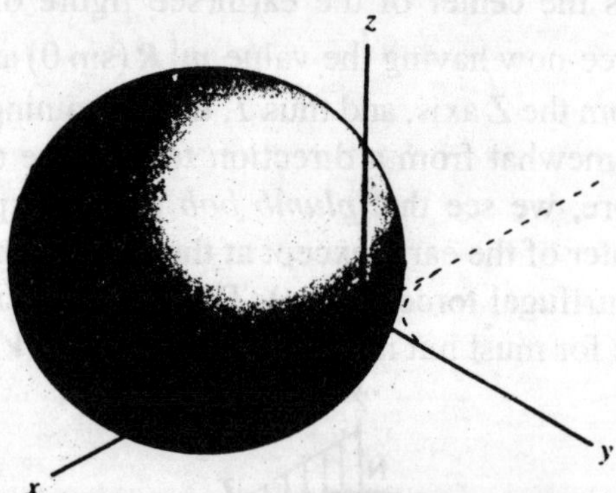

Figure 6.20 : *Primary Coriolis effect of free.*

Now, let us consider a body which is held above the earth's surface in such a way that is would always be above the same point on the earth's surface.. (The body thus moves with the earth with the same angular motion). If the body is released, it is called a free fall. The body will attain initially a downward velocity V_{xyz} relative to the earth's surface (see figure 6.18). Now in addition to a centrifugal force described earlier, we have a *Coriolid force* given as

$$F_{\text{coriolis}} = -2m\omega \times V_{xyz}$$

In figure 6.19 vector ω and V_{xyz} are shown. Coriolis force must point of the right. (do not forget the minus sign). If a mass is dropped from a position *xyz* above a target then as a result of coriolis force, the mass would slightly curve away from the target (see Figure 6.20) even if there were no friction, wind etc., to complicate matters. Furthermore, the induced motion in the *x* direction itself induces Coriolis force componentss of a smaller order in the *y* direction, an so forth. You can now, surely appreciate how difficult a "free fall" can really become when great accuracy is attempted.

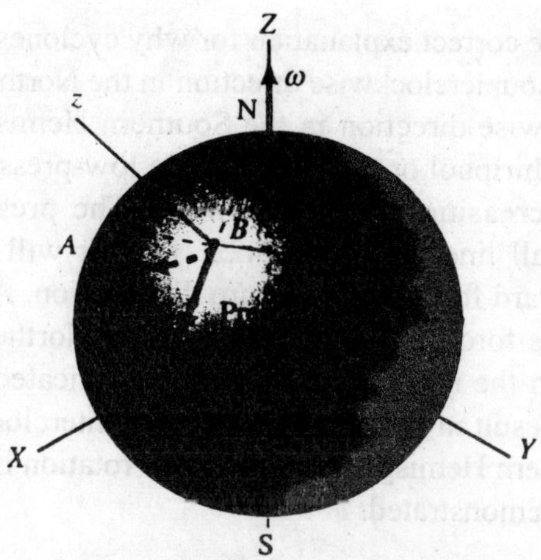

Figure 6.21 : *Corilolis effect of wind.*

Finally, consider a current of air or a current of water moving in the Northern Hemisphere. The fluid would move in the direction of the pressure drop in the absence of corioli's force. The pressure drop has been shown for simplicity along a meridian line pointing toward the equator, in figure 6.21. For fluid motion in this direction, a Coriolis force will be present in the negative *y*

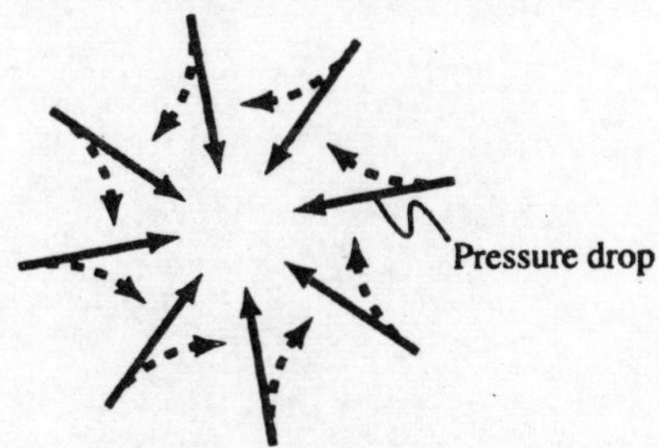

Figure 6.22 : *Beginning of whirlpool.*

direction and so the fluid will follow the dashed-line path *BA*. The prime induced motion is to thr right of the direction of flow developed by the pressure alone. Similarly, in the Southern Hemisphere, the Coriolis force induces a motion to the left of the flow that would be present under the action of the pressure drop alone. Such effects are of significance in meteorology and oceanography.

This is the correct explanation for why cyclones and whirlpools rotates in a counterclockwise direction in the Northern Hemisphere and a clockwise direction in the Southern Hemisphere. In order to start, a whirlpool or cyclone needs a low-pressure region with pressure increasing radially outward. The pressure drops are shown as full lines in figure 6.22. The air will begin to move radially inward for such a pressure distribution. As this happens, The Coriolis force causes the fluid in the Northern Hemisphere to swerve to the right of its motion, as indicated by the dashed lines. This result in the beginning of a counterclockwise motion.. In the Southern Hemisphere a clockwise rotation is induced, it can be readily demonstrated.

7

Kinetics of Plane Motion of Rigid Bodies

Introduction

From the previous chapters we have learned that in kinematics the motio of a rigid body at any time *t* can be considered to a superposition of a translational motion and a rotaitonal motion. The translational motion may have the aclual instantaneous velocity of any point, and the angular velocity of the rotation, ω, then has its axis of rotation through the chosen point. The center of mass of the rigid body is the most convenient point. The translatory motion can then be found form particle dynamics. You will recall that the motion of the center of mass of any aggregate of particles (this includes a rigid body) is related to the total external force by the equation

$$\boldsymbol{F} = M\dot{\boldsymbol{V}}_c \tag{7.1}$$

where *M* is the total mass of the aggregate. On Integration, we obtain the equation for the motion of the center of mass.

$$\boldsymbol{M}_A = \dot{\boldsymbol{H}}_A \tag{7.2}$$

for any system of particles where the point *A* about which moments of force and linear momentum are to be taken can be (1) the mass center, a point fixed in an inertial reference, (2) a point accelerating toward or away from the mass center. For these points, we shall later show that when above equation is applied to rigid bodies the angular velocity vector is involved along with inertia tensor. After we find the motion of the mass center from

Equation 7.1 and the angular velocity ω from Equation 7.2, we get the instantanious motion by letting the entire body have the velocity V_c plus the angular velocity ω, with the axis of rotation going through the center of mass.

Moment-of-Momentum Equations

Consider a rigid body whose each particle moves parallel to a plane, such a body is said to be in plane motion relative to this plane. We shall considers that axes *XY* are in the aforementioned plane in the ensuing discussion. The *Z* axis axis is then normal to the velocity vector of each point in the body. Furather, we consider only the situation where *XYZ* is an *inertial referecne.* Figure 7.1 shows a body undergoing plane motion relative to *XYZ* as described above.

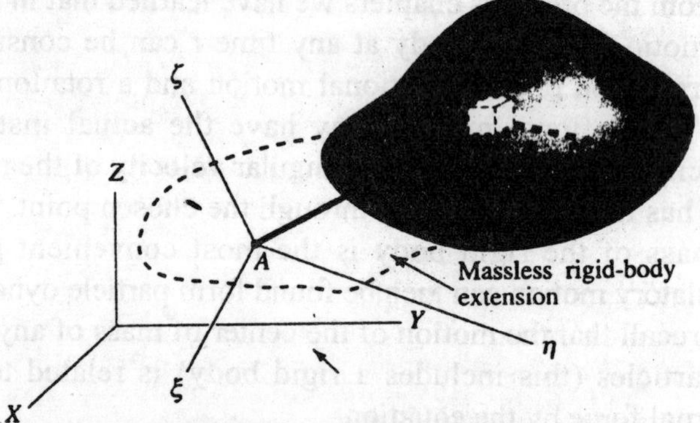

Figure 7.1 : *Body undergoing plane motion parallel to XY plane ξηζ translate with* A.

Now, choose a point *A* which is either a part of this or a hypothetical mass less rigid body extension of this body. A position ρ from *A* shows as element *dm* of the body. The velocity **V'** of *dm* relative to *A* is simply the velocity. of *dm* relative to any reference ξηζ, which translate with *A* relative to *XYZ*. Similarly, the linear momentum of *dm* relative to *A* (i.e., **V'** *dm*) is the linear momentum of *dm* relative to ξηζ translating with *A*. Now the moment of this momentum (i.e. the angular momentum) dH_A about *A* can be given as :-

$$dH_A = \rho \times V'dm = \rho \times \left(\frac{d\rho}{dt}\right)_{\xi\eta\zeta} dm$$

But as A is fixed in the body (or in a hypothetical massless extension of the body) and dm is a part of the body having mass, the vector ρ must be *fixed* in the body and , accordingly,

$$\left(\frac{d\rho}{dt}\right)_{\xi\eta\zeta} = \omega \times \rho$$

where ω is the angular velocity of the body relative of ξηζ. However, since ξηζ translates relative to XYZ, ω is the angular velocity of the body relative to XYZ as well. Hence, we can say :

$$d\boldsymbol{H}_A = \rho \times (\omega \times \rho)\, dm \qquad (7.3)$$

The angular velocity ω for the plane motion relative to the XY plane must have a direction normal to the XY plane.

Reference ξηζ has helped us in obtaining equation 7.3 and now we don't need it any longer so we can disperse it. Instead we *fix* reference xyz to the body at point A such that the z axis is normal to the plane of motion while the other two axes have arbitrary orientations normal to z (see figure 7.2)

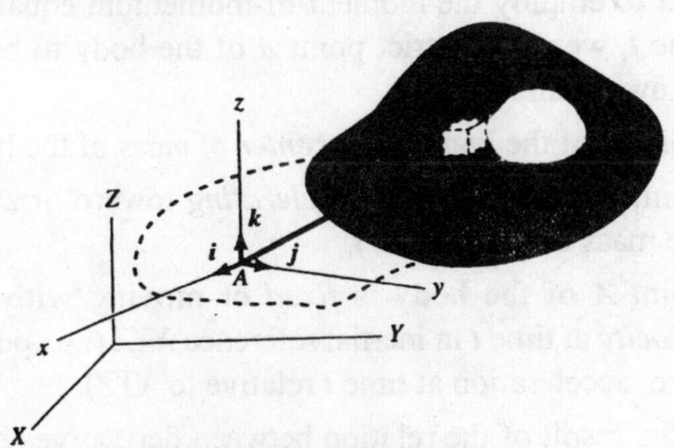

Figure 7.2 : *Reference xyz fixed to body at A; reference XYZ is inertial*

Because of the plane motion restriction the z axis will remain normal to XY, as the body moves. Next, we evaluate Equation 7.3 in terms of components relative of xyz as follows :

$$(dH_A)_x\, \boldsymbol{i} + (dH_A)_y\, \boldsymbol{j} + (dH_A)_z\, \boldsymbol{k}$$
$$= (s\boldsymbol{i} + y\boldsymbol{i} + z\boldsymbol{k}) + [(\mathbf{w}\boldsymbol{K}) \times (x\boldsymbol{i} + \mathrm{y}\mathbf{j} + \mathrm{z}\boldsymbol{k})]dm$$

The scalar equations which result form the foregoing vector equations are

$$(dH_A)_x = -\omega xz\, dm$$

$$(dH_A)_y = -\omega yz\, dm$$

$$(dH_a)_z = \omega\,(x^2 + y^2)\, dm$$

Integrating we get

$$(H_A)_x = -\iiint_M \omega xz\, dm = -\omega \iiint_M xz\, dm = -\omega I_{xz}$$

$$(H_A)_y = -\iiint_M \omega yz\, dm = -\omega \iiint_M yz\, dm = -\omega I_{yz} \qquad (7.4)$$

$$(H_A)_z = -\iiint_M \omega(x^2 + y^2)\, dm = -\omega \iiint_M (x^2 + y^2)\, dm = \omega I_{zz}$$

Now, we have the angular momentum componets for reference *xyz* at *A*. Note that, the inertia terms I_{xz}, *Iyz*, I_{zz} must be constant, because *xyz* is fixed to the body.

In order to employ the moment-of-momentum equation $\boldsymbol{M}_A = \boldsymbol{H}_A$ at time *t*, we next restrict point *A* of the body to be any one of the following three cases.

1. Point *A* of the body is the *center of mass* of the body.
2. Point *A* of the body is *accelerating* toward or away form the mass center at time *t*.
3. Point *A* of the body is *fixed* or moving with *constant velocity* at time *t* in inertial reference *XYZ* (i.e., point *A* has zero acceleration at time *t* relative to *XYZ*).

Using the result of the relation between derivatives of vectors as seen from different references, next we can say, using references *XYZ* and *xyz*:

$$\left(\frac{d\boldsymbol{H}_A}{dt}\right)_{XYZ} = \left(\frac{d\boldsymbol{H}_A}{dt}\right)_{xyz} + \omega \times \boldsymbol{H}_A$$

where ω is the angular velocity of *xyz* and thus of the body relative to *XYZ*. Hence, the *moment-of-momentum* equation can be written as follows :

$$M_A = \left(\frac{dH_A}{dt}\right)_{xyz} + \omega \times H_A \qquad (7.5)$$

Using Equation 7.4 for the components of H_A we get this equation:

$$M_A = \frac{d}{dt_{xyz}}\left(-\omega I_{xz} i - \omega I_{yz} j + \omega I_{zz} k\right)$$

$$+ \omega k \times \left(-\omega I_{xz} i - \omega I_{yz} j + \omega I_{zz} k\right)$$

Noting that ***i, j,*** and ***k*** are constant vectors as seen from *xyz* as are the inertia therms, we obtain

$$M_A = -\dot{\omega} I_{xz}\, i - \dot{\omega}\, l_{yz} j + \dot{\omega}\, l_{zz} k - \dot{\omega}\, l_{xz} j + \dot{\omega}\, l_{xz} j + \dot{\omega}\, l_{yz}\, i$$

The scalar forms of the above equation ar then

$$(M_A)_x = -l_{xz}\dot{\omega} + l_{yz}\dot{\omega} \qquad (7.6\ a)$$

$$(M_A)_y = -l_{yz}\dot{\omega} - l_{xz}\dot{\omega} \qquad (7.6\ b)$$

$$(M_A)_z = l_{zz}\dot{\omega} \qquad (7.6\ c)$$

With emphasis it should be noted that the angular velocity as given by ω (and later by $\dot{\omega}$) is always taken *relative to the inertial reference XYZ,* whereas the moments of forces (as given by $(M_A)_x$, $(M_A)_y$, and $(M_A)_z$) as well as the inertia tensos components are always taken about the axes *xyz* fixed to the body at *A* Equation (7.6) Equation (7.6) are the *general angular momentum equations for plane motion.* The last equation is quiet familiar to you form your work in physics. There you expressed it as

$$T = I\alpha \qquad (7.7)$$

or as

$$T = I\ddot{\theta} \qquad (7.8)$$

Now, we shall consider special cases of plane motion, starting with the most simple case and going toward the most general case. However, always rememeber that for all plane motions realtives to an inertial reference, the moment of the forces about the *z* axia at *A always* equals $l_{zz}\dot{\omega}$ The other two equations of 7.6 may get simplified for various special plane motions.

Also, to use Equation 7.6, we must remember that point *A* is a part of the body (because we took ρ to be fix in the body so we could use $d\rho/dt = \omega \times \rho$). Furthermore, to render the inertia tensor componets constants the *xyz* axes are fixed to the body.

Finally, ω, θ and their derilvatives are measured from the inertial reference *XYZ* as noted from the derivative.

Pure Rotation of a body of revolution about its axis of revolution

In Figure 7.3 a uniform body of revolution is shown. We have plane motion parallel to any plane for which the axis of revolution is a normal when the body is undergoing pure rotation about the axis of revolution fixed in inertial space reference *XYZ*. A reference *xyz* is *fixed* to the body such that *z* axis is collinear with the axis of revolution. We can choose for the origin of reference *xyz* any point *A* along this axis since all point along the axis of revolution are fixed in inertial space *XYZ*. The *x* and *y* axes forming a right handed triad than have arbitrary orientation. For simplicity, we choose *xyz* collinera with axes *XYZ* at time *t*. Clearly, the plane *zy* is a plane of symmetry for this body, and the *x* axis is normal to this plane of symmetry. As a consequence, $I_{xy} = I_{xz} = 0$.

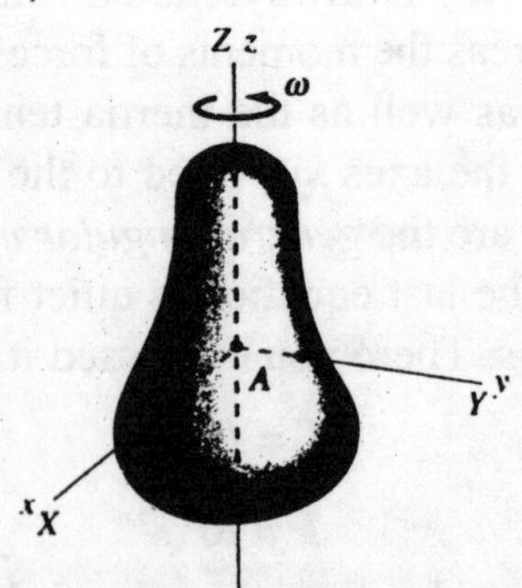

Figure 7.3 : *Rigid uniform body of revolution at time t.*

Similarly we conclude with *y* normal to a plane of symmetry, *xz*, that $I_{yx} = I_{yz} - 0$. Hence, *xyz* are principal axes. Returning to Equation 7.6 we find that only one equation of the set has nonzero moment, and that is the familiar equation

$$M_z = I_{zz}\dot{\omega}_z \tag{7.9}$$

The other pair of equations from 7.6 yield

$$M_x = 0$$

$$M_y = 0 \qquad (7.10)$$

We now turn to *Newton's law*. For this we must use the inertial reference *XYZ*. Note that the axes *xyz* and axes *XYZ* have been taken as collinear at the instant *t* shown in figure 7.3. This means the directions of *X* and *X* are the same at the same at his instant and it is only the direction that is significant here so the forces on the body needed in *Newton's law* such as F_x can be denoted by F_x . Since the center of mass of the body is stationary at all times (it is on the axis or rotation). Then form *Newton's law* we can say that :

$$\sum F_x = 0$$

$$\sum F_y = 0 \qquad (7.11)$$

$$\sum F_z = 0$$

Thus, the applied forces at any time *t*, the supporting forces, and the weight of the body of revolution safisfy *all* the equations of equilibrium *except* for meotion about the axis of revolution where Equation 7.9 applies.

Note, that the key equation (7.9) has the *same form* as Newton's law for *rectilinear translation* of a particle along as axis, say the *x* axis. We write body equations together as follows :

$$M_z = I_{zz}\ddot{\theta} \qquad (7.12a)$$

$$F_X = M\ddot{X} \qquad (7.12b)$$

Pure rotation of a body with two Orthogonal planes of symmetry

In Figure 7.4, a uniform body having two rothogonal planes of symmetry is shown, where in (a) we have shown the mentioned planes of symmetry and in (b) we have shown a view along the intersection of the planes of symmetry. We shall consider pure rotation of such a body about which we take as the *z* axis. As shown in the diagram, the origin A can be taken anywhere along the axis of rotation, and x and y axes are fixed in the planes of symmetry. XYZ is taken collinear with xyz at time t. The identical equations apply to this case as to the previous case of a body of revolution.

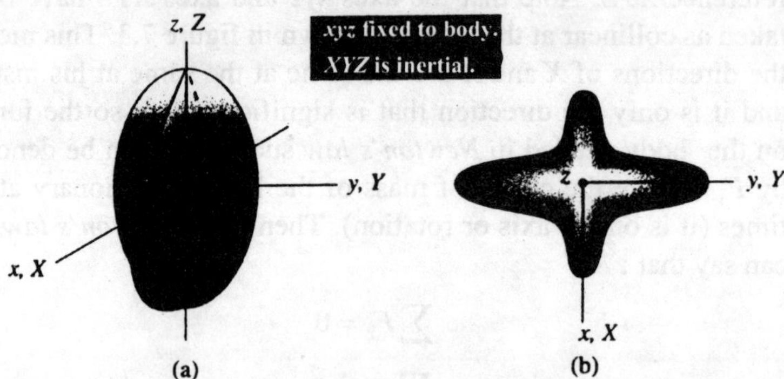

Figure 7.4 : *Body with two orghogonal planes of symmetry at time t.*

The torque is the next example is a function of angular speed.

Pure Rotation of Slabike bodies

Now, we consider bodies that have a *single* plane of symmetry. such bodies are called as slab-like bodies as shown if Figure 7.5. The body is oriented such that the plane of symmetry is parallel to the XY plane. We shall now consider the pure rotation of such a body about a foxed axis normal to the XY plane and going through a point A in the plane of symmetry of the body. We fis a reference xyz to the body at point A with xy in the plane os symmetry and z along the axis of rotation.

The angular velocity ω is then along the z axis. It is clear immediately that $l_{zx} = l_{zy} = 0$ as z is normal to the plane of symmetry. And so the *moment of momentum* equations become for this case:

$$M_x = 0$$

$$M_y = 0$$

$$M_z = l_{zz}\dot{\omega}_z \qquad (7.13)$$

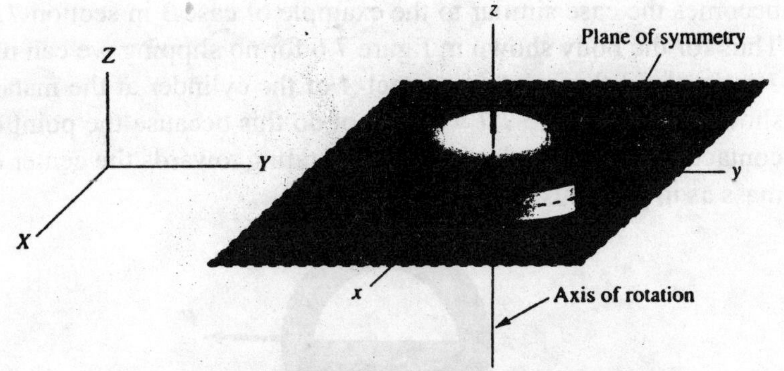

Figure 7.5 : *Slablike body undergoing pure rotation.*

We do not have equilibrium condition for the center of mass if the center of mass is not at a position along the axis of rotation. It will be undergoing circular motion. However, through *Newton's law* we can relate the external forces on the body to the acceleration of the mass center. Te yield enough equations to solve the problem, we have to use the kinematics of rigid body motion.

Rolling Slabike Bodies

Now we consider the rolling without slipping of slabike bodies such as cylinder, plane gears or spheres. The point of contact of the body has instantaneously zeor velocity and so we have pure instantaneous rotation about this point of contact. For getting velocities of points on such a rolling body, we can imagine that there is a hinge at the point of contact. Also by using the simple formulate *RQ* we can compute the acceleration of the center of a rolling without slipping sphere or cylinder. Finally if the angular speed is zero, we can find the acceeleration of any point in the cylinder of sphere by again imagining a hinge at the point of contact.

An important fact for the cylinders and spheres is that for rolling without slipping the acceleratin of the contact p oint on the cylinder or sphere is toward the geometric center of the cylinder or sphere. If the center of mass of the body lies along the line *AO* from the conatact point *A* to the geometric center *O* thaen we use equation 7.6 die oiubt *A*. This is justified as it becomes the case similar to the example of case 3 in section 7.2. Thus for the body shown in figure 7.6 for no slipping we can use $T = I\alpha$ about the point of contact A of the cylinder at the instant shown. But in figure 7.7 we can not do this because the point of contact A of the cylinder is not accelerating towards the center of mass as in previous case.

Figure 7.6 : *Point* A *accelerates toward center of mass.*

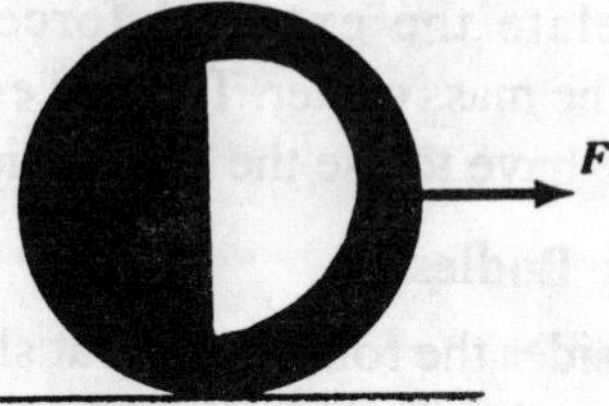

Figure 7.7 : *Point* A *does not accelerate toward center.*

We shall now examine a problem involving rolling without slipping. The equations of motion, which are deduce, are the same in the previous section.

General plane motion of a Slablike body

Now, we consider *general plane motion* of slablike bodies. The motion which is to be studied is parallel to the plane of symmetry. Accordingly, we use the center of mass. The angular velocity vector ω will be normal to the plane of symmetry, and, in accordance with Chasles' theorem, well be taken to pass through the center of mass. The translational velocity vector V_c will be parallel to the plane of symmetry. The figure 7.8 shows that a reference *xyz* is fixed at the center of mass of the body such that the *xy* plane coincides with the plane of symmetry. The inertial reference is parallel to *xyz* at time *t,* the actual instantanious axis of rotation is also shown. For the same conditions used in sectin 7.5 for slablike bodies, the *moment -of -momentum* equations become

$$M_x = 0 \quad (7.14a)$$

$$M_y = 0 \quad (7.14b)$$

$$M_z = 0 \quad (7.14c)$$

Furthermore, we must have *equilibrium* in the *z* direction in accordance with the center of mass while the full form of *Newton's law* holds in the *x* and *y* directions.

$$\sum F_z = 0 \quad (7.15)$$

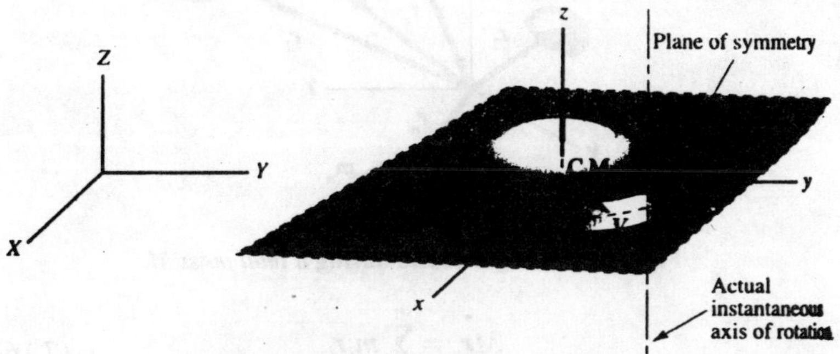

Figure 7.8 : *Slablike body undergoing general motion parallel to* XY.

As in section 7.5 we can generalize the results of this section to include the plane motion of a body having at the center of *mass in principal* axis *z* normal to the plane of motion *XY*. We get the same equations of motions as for the slablike body for principal axis *xyz* at *A*.

Pure rotation of an Arbitrary rigid body

Now, we consider a body rotation about an axis of rotation fixed in inertial space. This body is having an arbitrary distribution of mass. We consider this axis to the *z* axis fixed in the body as well as being an inertial coordinate axis *Z*. We can take the origin of *xyz* anywhere along the z axis since all such point are fixed in inertial space. Since I_{zx} and I_{zy} are not equal to zero so the moment of momentum equation to be used will now be the general equation 7.6. There will be no acceleration, if the center of mass is along the *z* axis. We can then apply thr rules of statics to the center of mass. For other cases we shall often need to use *Newton's law* for the center of mass. In this regard it will be helpful to note from the definition of the center of mass that for a system of rigid bodies such as is shown in Figure 7.9.

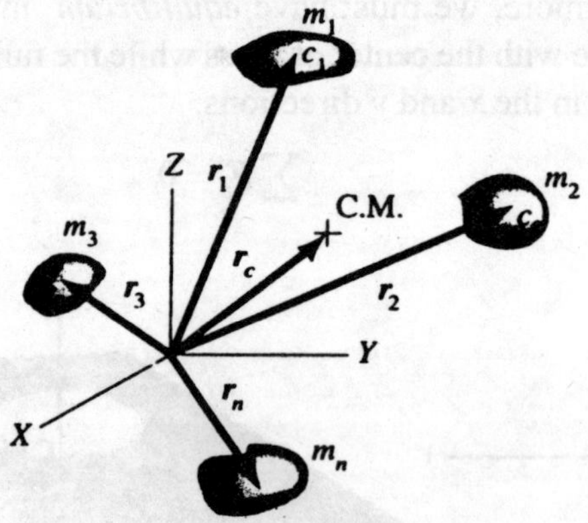

Figure 7.9 : *n rigid bodies having a total mass M.*

$$M\boldsymbol{r}_c = \sum_{i=1}^{n} m_i \boldsymbol{r}_i \tag{7.16}$$

where m_i is the mass of the *i*th rigid body,***ri*** is the position vector to the center of mass of the *i*th rigid body, *M* is the total mass and $\boldsymbol{r}_c$ is the postion vector to the center of mass of the system. We can then say on differenatiating:

$$M\dot{\boldsymbol{r}}_c = \sum_{i=1}^{n} m_i \dot{\boldsymbol{r}}_i \tag{7.17}$$

$$M\ddot{\boldsymbol{r}}_c = \sum_{i=1}^{n} m_i \ddot{\boldsymbol{r}}_i \tag{7.18}$$

We can draw the conclusion that in *Newton's law* for the mass center of a system of rigid bodies, that we can use the centers of mass of the component parts of the system as given on the right side of Equation 7.18 rather than the center of mass of the total mass.

Balancing

We shall now set the criteria for the condition of dynamic balance in a rotating body. Then we have certain requirement for achieving balance in a rotating body. Consider some arbitrary rigid body rotating with angular speed ω and a rate of change of angular speed to $\dot{\omega}$ about axis *AB,* as shown in figure 7.10. We shall set up a general equations for determining the supporting forces at the bearings. Consider point *a* on the axis of rotation at the bearing *A* and establish a set of axes *xyz* fixed to the rotation body with the *z* axis corresponding to the axis of rotation. For convenience we choose the axes *x* and *y*. Axes *XYZ* are, as usual, inertial axes. Using the *moment-of-momentum* equations (a) and (b) in Equation 7.6 and including only *dynamic* forces, we get for point *a:*

$$B_y l = -I_{xz}\dot{\omega} + I_{yz}\omega^2 \tag{7.19a}$$

$$B_x l = -I_{yz}\dot{\omega} + I_{xz}\omega^2 \tag{7.19b}$$

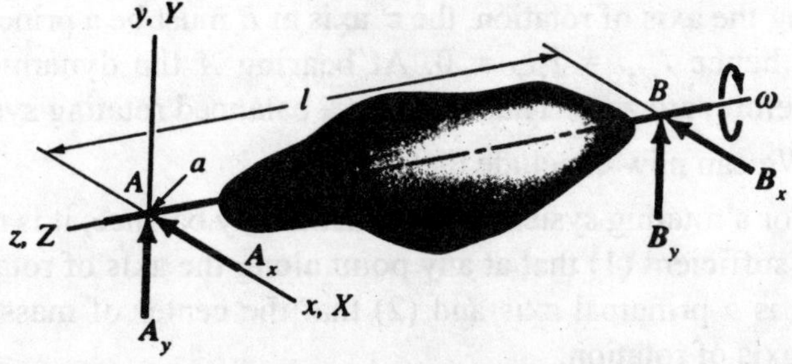

Figure 7.10 : *Rotating body.*

The products of inertia $I_{x'z'}$ are $I_{y'z'}$ are zero. Now the *xy* axes and hence the *x'y'* axes can have any orientation as long as they are normal the axis of rotation. This menas that at the *E* the *z'* axi yiels a zero product of inertia for all axes normal to it. *Z'* is a principal axis for point *E*. And since *E* is any point on the axis of rotation, we can say the following:

If the axis of rotation is a principal axis at any point along this axis and if the center of mass is on the axis of rotation, then the axis of rotation is a principal axis at all points on it.

We now consider Figure 7.11, where reference $x'y'z'$ is set up at bearing B. We can next employ the **moment-of-momentum** equation (7.6) for these axes at B. We get

$$-A_y l = -I_{x'z'}\dot{\omega} + I_{y'z'}\omega^2$$

$$-A_x l = -I_{y'z'}\dot{\omega} - I_{x'z'}\omega^2$$

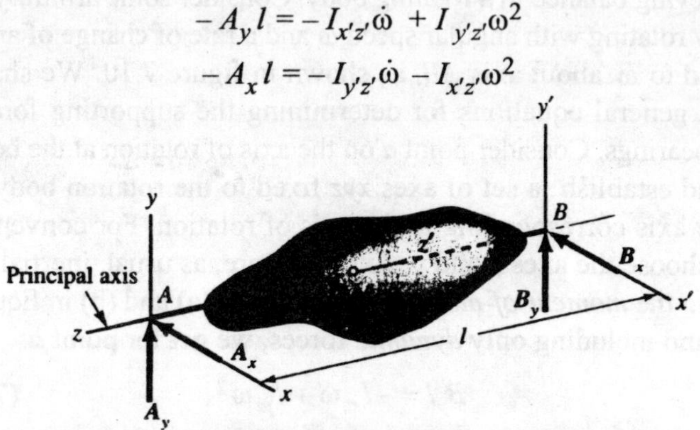

Figure 7.11 : *Reference x'y'z' fixed at B.*

With the z axis a principal axis at A and the center of mass along the axis of rotation, the z' axis at B must be a principal axis, and hence $I_{y'z'} = I_{x'z'} = 0$. At bearing A the dynamic forces, therefore, are zero. Thus we have a balanced rotating system.

We can now conclude that:

For a rotating system to be dynamically balance, it is necessary and sufficient (1) that at any point along the axis of rotation thes axis is a principal axis and (2) that the center of mass is along the axis of rotation.

Laws for Rotary Motion

First Law

According the the I law, a body continues in its state of rest or of rotation about an axis with constant angular velocity unless it is compelled by an external torque to change the state.

But, in actual practice, we see that when a body is rotating about an axis with constant angular velocity, it comes to rest after some time. This is due to resistance and friction between the body and its bearing on axis. If these forces had been absent, the body would have gone or rotaing indefinitely.

Second Law

It states that the rate of change of angular momentum of a rotating body proportional to the external torque applied on the body and takes plade in the direction of the torque.

Consider a body of moment of inertia I rotating with an angular velocity ω_0. It is acted upon a torque T and the angular velocity of the body becomes ω. Then we have

ω_0 = Intial angular velocity,

ω_0 = Final angular velocity,

I = Moment of Inertia of the body = Mk^2

M = Mass of the body and k = radius of gyration. At this radius the whole mass of the body is assumed to concentrate in rotation.

T = Torque acting on the body,

t = time in seconds.

Initial angular momentum of the body

= Moment of Inertia × Initial angular velocity = $I \times \omega_0 = I\omega_0$

Final angular momentum

$$= I \times \omega = I\omega$$

∴ Change of angular momentum

= Final angular momentum – Initial angular moments

$= I\omega - I\omega_0 = I(\omega - \omega_0)$

∴ Rate of change of angular momentum

$$= \frac{\text{Change of momentum}}{\text{Time}} = \frac{I(\omega - \omega_0)}{t}$$

But we know that,

$$\frac{(\omega - \omega_0)}{t} = \alpha = \text{Angular acceleration}$$

∴ Rate of change of angular momentum = $I \times \alpha = I\alpha$.

But, the rate of change of angular momentum is directly proportional to the external torque applied on the body as stated by the second law of motion for rotation.

$\therefore$ Torque $\propto$ Rate of change of angular momentum

or $T \propto I\alpha$ $\quad\therefore\quad T = KI\alpha$

where k is a constant of propertionality.

Unit torque is that which produces unit angular acceleration on unit moment of inertia, then we have $T = 1$, $I = 1$, $\alpha = 1$. Substituting these values in equation (*i*), we get

$$l = k \times 1 \times 1 = k.$$

Substituting this value of $k = 1$ in equation (*i*), we get

$$T = I\alpha \tag{7.20}$$

Units of Torque:

The unit of torque depends upon the unit of moment of inertial I and $I = mk^2$, k is the radius of gyration from the axis of rotation. It is the radius of gyration where the whole mass of the rotation body is supposed to be concentrated. If I is in $kg\ m^2$, then T is in Newton-meter written as N m.

Note. If the value of M is taken in *kg*, then the unit of torque is in *N m*. But if the value of M is taken as, where W is in *kgf*, then unit of torque will be in *kgf-m*.

Third Law

According to this law to every torque there is always an equal and opposite torque.

Kinetic Energy due to Rotation

Figure 7.12 shows a rigid body rotation about O, Now,

Let ω = Angular velocity of the body

dm = Elementary mass of the body

r = Radius of elementary from O

V = Tangential velocity of elementary mass

$= \omega \times r$

K.E. of the elementary mass

$$= \frac{1}{2} \times \text{mass} \times \text{velocity}^2 = \frac{1}{2} \times dm \times V^2.$$

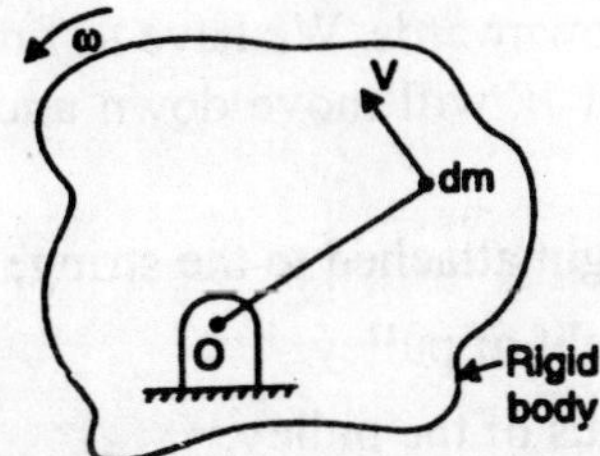

Figure 7.12 : *Rigid body rotating about O.*

By integrating the above equation we obtain the K.E. of the whole body. Hence K.E. of the body

$$= \int \frac{1}{2} dm \times V^2 = \frac{1}{2} \int dm \times (\omega r)^2 \qquad (\because V = \omega r)$$

$$= \frac{1}{2} \int \omega^2 r^2 dm = \frac{1}{2} \omega^2 \int r^2 dm. \qquad (\because \omega \text{ is a constant})$$

But $\int r^2\, dm = I$ = Moment of Inertia of the body about O.

$$\therefore \text{K.E. of the body} = \frac{1}{2} \omega^2 I. \qquad (7.21)$$

If mass is taken in kg, the unit of K.E. will be Nm.

Total Kinetic Energy of a body

Equation (7.21) gives the K.E. of the body due to rotation only. But if the body possesses both the motion of translation as well as the motion of rotation, then the total kinetic energy of body is equal to the K.E. due to rotation plus K.E. due to translation. If the linear velocity of the body is V, then K.E. due to linear velocity $= \frac{1}{2} mV^2$.

$$\therefore \text{ Total Kinetic Energy} = \frac{1}{2} I\omega^2 + \frac{1}{2} mv^2. \qquad (7.22)$$

Rotation due to a Weight W attached to one end of a string passing over a pulley of weight W_0

Consider a weight W attached to one end of an inextensible string passing over a pulley. The other end of the string in attached to the periphery of the pulley as shown in figure 7.13. Let the weight W is moving downwards. We have to find the acceleration with which the weight W will move down and the tension P in the string.

Let W = Weight attached to the string,

W_0 = Weight of pulley,

R = Radius of the pulley,

P = Tension in string,

a = Acceleration with which the weight W moves down,

I = Moment of Inertia of the pulley about the axis of rotation, and

α = Angular acceleration of the pulley.

Net force on this weight is act downwards as the weight W is moving downward with an acceleration f

$\therefore$ Net downward force on the weight $W = (W - P)$.

Using, net force = mass × acceleration

$$\therefore \quad W - P = \frac{W}{g} \times a \qquad \left(\because \text{Mass} = \frac{\text{Weight}}{g}\right) \qquad (i)$$

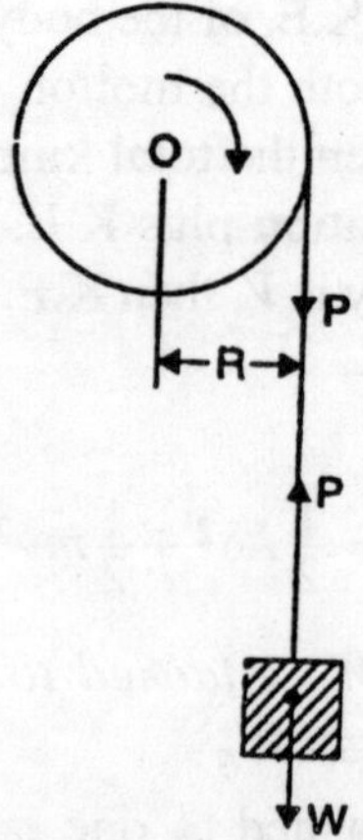

Figure 7.13

Now consider the motion of pulley

By equation (7.20) the torque on pulley is given as :

$$T = I \times \alpha$$

where I = Moment of inertia of the pulley about the axis of rotation.

If pulley is considered as solid disc, then moment of inertia of pulley is given by equation

$$\therefore \qquad I = \frac{MR^2}{2} \qquad (\because \text{ Solid disc is like cylinder})$$

In the above equation M = Mass of the pulley

$$= \frac{\text{Weight of the pulley}}{g} = \frac{W_0}{g}$$

$$I = \frac{W_0}{g} \times \frac{R^2}{2}$$

On substitution of I in equation in (*ii*), we obtain

$$T = \left(\frac{W_0}{g} \times \frac{R^2}{2}\right) \times \alpha.$$

But the relation between angular acceleration (α) and linear acceleration is given by

$$a = \alpha \times R$$

$$\therefore \qquad \alpha = \frac{a}{R}.$$

Substituting the value of α in equation (*iii*), we obtain

$$T = \left(\frac{W_0}{g} \times \frac{R^2}{2}\right) \times \frac{a}{R}$$

But the torque on the pulley is due to the tension P and its magnitude is given by

$$T = P \times R.$$

Substituting this value of T in equation (*iii*), we get

$$P \times R = \left(\frac{W_0}{g} \times \frac{R^2}{2}\right) \times \frac{a}{R} \quad \text{or} \quad P = \frac{W_0}{2g} \times a.$$

Adding equations (*i*) and (*v*), we get

$$W = \frac{W}{g} \times a + \frac{W_0}{2g} \times a = \frac{a}{g}\left[W + \frac{W_0}{2}\right]$$

$$\therefore \quad a = \frac{gW}{\left(W + \frac{W_0}{2}\right)} \text{ m/s}^2 \qquad (7.23)$$

To find the tension (*P*) in the string, the value of *a* is substituted in equation (*i*).

$$\therefore \quad W - P = \frac{W}{g} \times \frac{gW}{\left(W + \frac{W_0}{2}\right)} = \frac{W^2}{\left(W + \frac{W_0}{2}\right)}$$

$$\therefore \quad P = W - \frac{W^2}{W + \frac{W_0}{2}} = W - \frac{2W^2}{(2W + W_0)}$$

$$= \frac{2W^2 + WW_0 - 2W^2}{(2W + W_0)} = \frac{WW_0}{(2W + W_0)} \qquad (7.24)$$

Rotation due to weight attached to the two ends of a string, which passes over a rough pulley of weight W_0

The two weights *W* and W_2 attached to the two ends of a string, which passes over a rough pulley of radius *R* as shown if Figure 7.14. The tension on both sides of the string will not be same as the pulley is rought and have certain weight. If $W_1 > W_2$ the weight W_1 will move downwards whereas the weight W_2 will move upwards with the same acceleration.

Let a = Acceleration of the system;

T_1 = Tension in the string to which weight W_1 is attached ;

T_2 = Tension in the string to which weight W_2 is attached ;

R = Radius of the pulley ;

I = Moment of inertia of the pulley about the axis of rotation ;

α = Angular acceleration ; and

W_0 = Weight of the pulley.

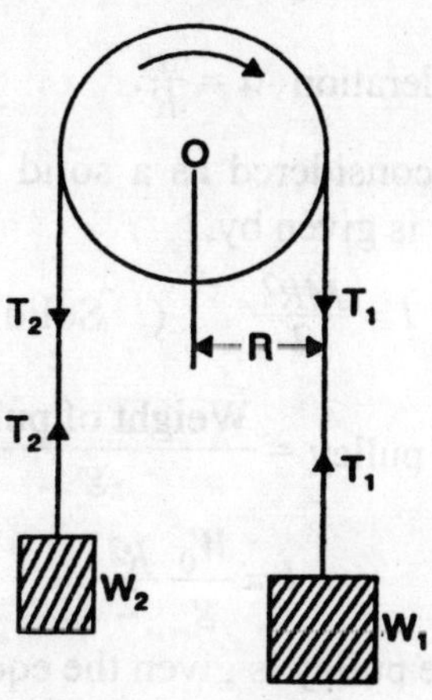

Figure 7.14

I. *Consider the motion of weight* W_1. The weight W_1 is moving downwards with an acceleration a.

$\therefore$ The net downwards force on weight $W = (W_1 - T_1)$

Mass of weight, $$W_1 = \frac{W_1}{g}.$$

Now using, net force = mass × acceleration

or $$(W_1 - T_1) = \frac{W_1}{g} \times a.$$

II. *Now consider the motion of weight* W_2.The weight W_2 is moving upwards with an acceleratin a

Net upwards force on weight $W_2 - (T_2 - W_2)$.

Mass of weight, $$W_2 = \frac{W_2}{g}$$

Using, net force = Mass × Acceleration

or $$(T_2 - W_2) = \frac{W_2}{g} \times a.$$

III. *Now consider the motion of the pulley*. The pulley is rotating with an angular acceleration, α.

But angular acceleration $\alpha = \frac{a}{R}$.

If the pulley is considered as a solid disc, then moment of inertia of the pulley is given by,

$$I = \frac{MR^2}{2} \qquad (\because \text{ Solid disc is like a cylinder})$$

where M = Mass of pulley $= \frac{\text{Weight of pulley}}{g} = \frac{W_0}{g}$.

$$\therefore \qquad I = \frac{W_0}{g}\,\frac{R^2}{2}.$$

The torque on the pulley is given the equation (7.20) as

$$T = I \times \alpha = \frac{W_0}{g}\,\frac{R^2}{2} \times \frac{a}{R} \qquad \left(\because \alpha = \frac{a}{R}\right) \qquad (iii)$$

But torque on the pulley = Torque due to T_1 – Torque due to T_2

$$= T_1 \times R - T_2 \times R = R\,[T_1 - T_2]$$

Substituting the value of torque in equation (*iii*), we get

$$R[T_1 - T_2] = \frac{W_0}{g}\,\frac{R^2}{2} \times \frac{a}{R}$$

$$\therefore \qquad T_1 - T_2 = \frac{W_0}{2g} \times a.$$

Adding equation (*i*), (*ii*), (*iv*), we get

$$W_1 - W_2 = \frac{W_1}{g} \times a + \frac{W_2}{g} \times a + \frac{W_0}{2g} \times a$$

$$= \frac{a}{g}\left[W_1 + W_2 + \frac{W_0}{2}\right]$$

$$\therefore \qquad a = \frac{g[W_1 - W_2]}{\left(W_1 + W_2 + \frac{W_0}{2}\right)} \qquad (7.25)$$

Substitutes the value of *a* in equation (*i*) and (*ii*) to find the tension T_1 and T_2.

Substituting the value of *a* in equation (*i*), we get

$$W_1 - T_1 = \frac{W_1}{g} \times \frac{g(W_1 - W_2)}{\left(W_1 + W_2 + \frac{W_0}{2}\right)} = \frac{W_1(W_1 - W_2)}{\left(W_1 + W_2 + \frac{W_0}{2}\right)}$$

$$\therefore\ T_1 = W - \frac{W_1(W_1 - W_2)}{\left(W_1 + W_2 + \frac{W_0}{2}\right)} = W_1\left[1 - \frac{(W_1 - W_2)}{\left(W_1 + W_2 + \frac{W_0}{2}\right)}\right]$$

$$= W_1\left[\frac{W_1 + W_2 + \frac{W_0}{2} - W_1 + W_2}{W_1 + W_2 + \frac{W_0}{2}}\right] = W_1\left[\frac{2W_2 + \frac{W_0}{2}}{W_1 + W_2 + \frac{W_0}{2}}\right] \quad (7.26)$$

Substituting the value of *a* in equation (*ii*), we get

$$T_2 - W_2 = \frac{W_2}{g} \times \frac{g(W_1 - W_2)}{\left(W_1 + W_2 + \frac{W_0}{2}\right)} = \frac{W_2(W_1 - W_2)}{\left(W_1 + W_2 + \frac{W_0}{2}\right)}$$

$$\therefore T_2 = W_2 + \frac{W_2(W_1 - W_2)}{\left(W_1 + W_2 + \frac{W_0}{2}\right)} = W_2\left[1 + \frac{W_1 - W_2}{\left(W_1 + W_2 + \frac{W_0}{2}\right)}\right]$$

$$= W_2\left[\frac{W_1 + W_2 + \frac{W_0}{2} + W_1 - W_2}{\left(W_1 + W_2 + \frac{W_0}{2}\right)}\right] = \frac{W_2\left[2W_1 + \frac{W_0}{2}\right]}{\left(W_1 + W_2 + \frac{W_0}{2}\right)} \quad (7.27)$$

Law of Conservation of Energy

It states that the energy can neither be created nor destroyed though it can be transformed from one form to another form. The second statement of this law is :

"The total energy possessed by a body remains constant provided no energy is added to or taken from it."

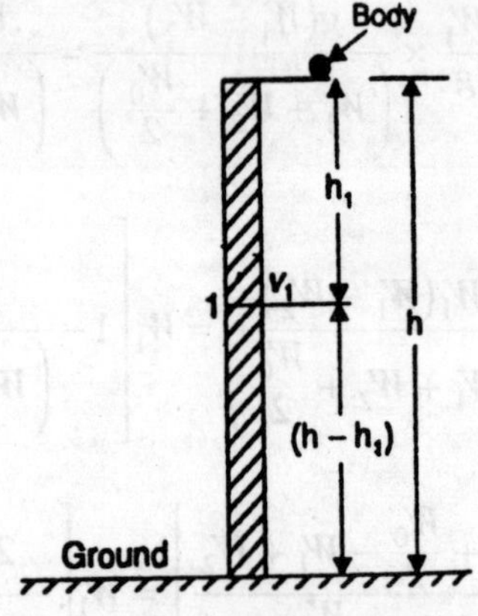

Figure 7.15

Figure 7.15 shows a body resting on the toop of a tower.

Let W = Weight of body

m = Mass of body $= \dfrac{W}{g}$

h = Height of tower

Now, let us find the potential energy and K.E. of the body at the top of the tower.

Potential energy of the body with respect to ground

$$= W \times h \text{ Nm} \qquad (i)$$

K.E. of the body

$$= \frac{1}{2} mv^2 = 0 \qquad (\because v = 0)$$

∴ Total energy of the top of tower

$$= \text{Potential energy} + \text{Kinetic energy}$$

$$= Wh + 0 = Wh$$

Suppose the body falls down by a height h_1 to the position 1 as shown in Figure 7.15. Let v_1 is the velocity at position 1.

Using equation $v^2 - u^2 = 2gh$, we get

$$v^2_1 - 0^2 = 2gh_1 \qquad (\because u = 0, h = h_1)$$

$$v_1^2 = 2gh_1$$

$\therefore$ K.E. of the body at postion 1

$$\frac{1}{2}mv_1^2 = \frac{1}{2}m \times 2gh_1 \qquad (\because v_1^2 = 2gh_1)$$

$$= mgh_1 = wh_1 \qquad (\because W = mg)$$

Potential energy at position 1

$= W \times$ Height of body at 1 *w.r.t* ground $= W \times (h - h_1)$

Total energy at position 1 =

= Potential energy + Kinetic energy

$= W(h - h_1) + Wh_1 = Wh - wh_1 + Wh_1 = Wh$

And it is the same energy, the body was possessing at the top of the tower.

Similarly, it can be proved that the total energy of the body on the ground will be equal to '*Wh*'.

Energy lost by a body falling on another body and to calculate the resistance offered the ground

Figure 7.16 shows a body of mass *M* falling on another body of mass *m*, from a height *h*. The mass *m*, before coming to rest penetrates into the ground through a distance S.

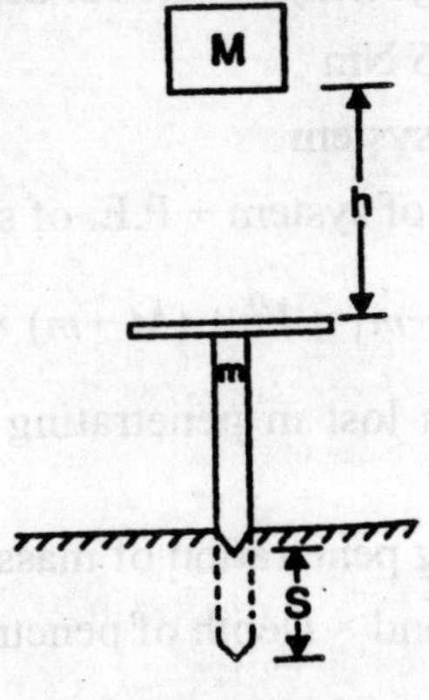

Figure 7.16

Let v = velocity of mass M before striking the mass m

$= \sqrt{2gh}$

V = Velocity of both masses M and m after impact (*i.e.*, common velocity)

We know that, the velocity of mass m before impact is zero. Applying law of conservation of momentum:

Total momentum of M and m before impact = Total momentum of both masses after impact

or $$M \times v + m \times 0 = (M + m) \times V$$

or $$M \times \sqrt{2gh} = (M + m) \times V$$

$$\therefore \quad V = \frac{M \times \sqrt{2gh}}{(M + m)} \qquad (i)$$

Resistance offered by the ground

The K.E. of the system (*i.e.* K.E. of both masses) just after the impact

$$= \frac{1}{2} \text{ (Total mass of system)} \times V^2$$

$$= \frac{1}{2} [\, M + m \,\} \times V^2 \text{ Nm}$$

and potential energy to the system just after the impact

= (Total weight of system) × Vertical depth of penetration

$= [\,(M + m) \times g] \times S$ Nm

Total energy of the system

= K.E. of system + P.E. of system

$$= \frac{1}{2} [M+m] \times V^2 + (M + m) \times g \times S$$

This total energy is lost in penetrating the mass m into the ground.

∴ Work done during penetration of mass m into ground

= Resistance of ground × Depth of penetration = $R \times S$

Now total energy lost is equal to work done during penetration.

$$\frac{1}{2}[M+m]\times V^2\,(M+m)\times g\times S=R\times S$$

By using the above equation, we can calculate the resistance *R*.

Impulse and Momentum

Let F = Net force acting on a rigid body in the direction of motion, through the C.G. of the body

m = Mass of the rigid body, and

a = acceleration of the body.

Then we have

$$\mathrm{F}=\mathrm{m}\times\mathrm{a}=\mathrm{m}\times\frac{dv}{dt}$$

($\because$ Acceleration is rate of change of velocity *i.e.* $a=\frac{dv}{dt}$)

or $$F\times dt=m\times dv$$

On, integrating the above equation, we obtain

$$\int_{t_1}^{t_2}F\times dt=\int_{v_1}^{v_2}m\times dv$$

$$=m\left(v_2-v_1\right)$$

It time-interval is very small $\int_{t_1}^{t_2}F\times dt$ is known as impulse. *Hence impule in the product of force and time when time is very small.* Momentum is the product of mass and velocity.

$$\therefore\ \text{impulse}=\int_{t_1}^{t_2}F\times dt=m\left(v_2-v_1\right)=mv_2-mv_1$$

Impule = Change of momentum

($\because$ Momentum = Mass × Velocity)

= Final momentum – Initial momentum

or Initial momentum + Impule = Final momentum

The above equation gives the relation beween impulse and momentum of a rigid body. The impulse-momentum equation is based on integrating the equation of motion with respect to time. The question of motion relatates force, velocity and time.

Note. (*i*) The unit of impulse is *Ns* (*i.e.* Newton second) in S.I. system.

(*ii*) The impulse-momentum approach is particularly convenient in situations when forces act for very small interval of time as in an impact or sudden blow.

(*iii*) In satellite motion, a combination of impule-momentum method and work-energy method is used.

Virtual work and principle of virtual work

Work done by a force on a body due to small imginary or virtual displacement of the body is known as virtual work. A virtual displacement is very small change in the position of a body. The body is not actuallty displaced, the displacement is only imagined, for which such displacement is called virtual displacement. The product of the formce and the virtual displacement in the direction of the force is called virtual work.

Work

The product of the force and distance is work. The distance is always in the direction of the forces. If a force is acting on a body and the body moves a distance *S* in the the direction of the force, as shown inGigure 7.17 then work done on the body is given by :

Work done = Force × Distance moved by the body in the direction of force.

$= P \times S$

But if the force acting of the body and the distance moved by the body are not in the same direction (as shown in figure 7.18), the work done will be given by :

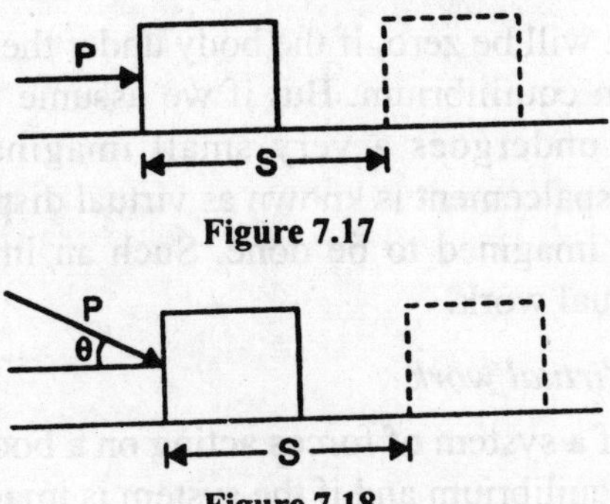

Figure 7.17

Figure 7.18

Work done = Component of the force in the direction of distance × Distance

$$= P \cos \theta \times S$$

where θ is the inclination of the force *P* with the direction of motion of the body.

It θ = 90°, cos θ = 90° = 0 and the work done will be zero. Hence if the angle between the line of action of the force land the direction of motion of the body is 90°, the work done will be zero.

(*i*) If the point of application of the force moves at right angles of the direction of the force, work done by a force is zero.

(*ii*) If the point of application of the force moves in the direction of the force, the work done by a force is said to be positive.

(*iii*) Work done by a force is said to be negative if the the point of application of the force moves in a direction opposite to the force.

(*iv*) The word virtual displacement is used to signify the fact that the displacement is not an actual one, it is only imagined. Hence a virtual displacement is an imaginary displacement which is possible but does not take place under action of forces.

Virtual work

The work done, by a force on a body due to small virtual (*i.e.*

imaginary) displacement of the body, is known as virtual work.

Work done will be zero, if the body under the action of system of forces is in equiliibrium. But if we assume that the body, in equilibrium, undergoes a very small imaginary displacemet (imaginary dispalcement is known as virtual displacement), some work will be imagined to be done. Such an imaginary work is known as virtual work.

Principle of Virtual work

It states, "If a system of forces acting on a body or a system of bodies be in equilibrium and if the system is imagined to undergo a displacement consistent with the geometrical conditions, then the algebraic sum of the virtual work done by the forces of the system is zero".

Work is the product of force and distance. In S.I. system, unit of force is Newton (N) whereas the unit of distance is in metren (m). Hence the unit of work in S.I. system is Newton-metre and written as N m.

The unit of force in M.K.S. system is Kilogram force (written as kgf) and unit of distance is metre (m). Hence unit of work in M.K.S. system is written as kgf-m.

One kgf-m = 9.81 N m.

Forces to be omitted while applying the principle of virtual work

The forces which may be omitted while applying the principle of virtual work are listed under :-

(*i*) Tension produced in an inextensible string. As the string is inextensible, its length does not undergo any change. Hence $\delta l = 0$ and $T \times \delta l = 0$.

(*ii*) Reactions between the body and any surface on which the body rolls without sliding.

(*iii*) Reactions at smooth pins and hinges.

(*iv*) The mutual action and reaction between the two bodies whose equilibrium is being considered togethere. As the action and reaction are equal and opposite, so the work done by them cancel each other.

(*v*). Forces normal to the direction of motion.

For most of the probles solved by virtual work method the system is assumed to be frictionless that is there is no friction in hinges, bearings or along slidding surfaces.

Important points to be remembered while applying the principle of virtual work

1. If all the forces act in the same direction, give the system a virtual displacement in the direction of the forces.
2. If the reaction R at a rigid support is to be calculated, as the rigid support is not supposed to move, the virtual work done by rigid sipport will be zero. Hence the reaction R will not appear in the equation of virtual work. To overcome this difficulty, remove the support and replace it by a force R. This force can be given a virtual displacement and the value of R will be obtained by applying principle of virtual work.
3. If the tension T in an inextensible string is to be calculated, replace the string by two forces each equal to T acting at the ends of the string. Now apply the priciple of virtual work and calculate the value of T.
4. If the roller moves horizontally, then the reacton of the roller dies not do any work. Also when the pin rotates, the reactions do not do any work.
5. Virtual displacements are denoted δx or $\delta\theta$ whereas actual displacements are denoted by dx or $d\theta$.

Uses of the principle of virtual work

The principle of virtual work is used in solving problems of statics in the following cases :

1. Beams 2. Framed structure 3. Lifting machine.

1. The principle of virtual work for problems on Beams. The principle of virtual work is applied for the problems on beams as given below :

Figure 7.19 shows a simple supported beam *AB*, which is hinged at *A*. Let a point load *P* acts at *C*. Then

R_A = Reactions at *A* acting upwards,

R_B = Reaction at *B* acting upwards,

L = Length of beam *AB*, and

a = Distance of point load from *A*.

Suppose keeping the point *A* is fixed the beam in given a virtual dispalcement at *B*.

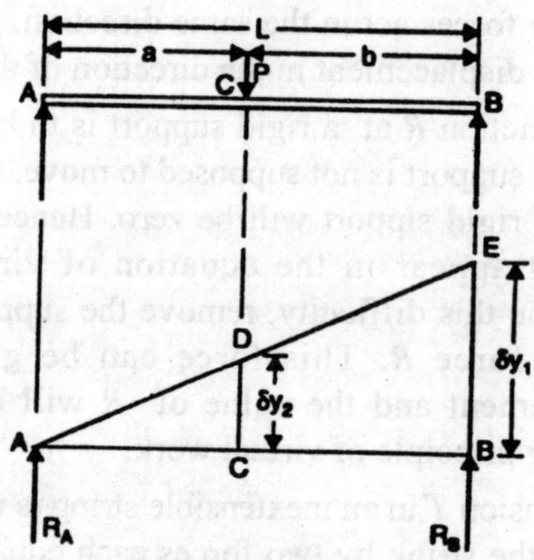

Figure 7.19

As the reaction R_B at *B* is acting upward, the virtaul displacement at *B* is assumed in the upward direction. The beam at *C* will also undergo an upward virtual displacement, but as *A* is assumed to the fixed, the upward virtual displacement is zero.

Let δy_1 = Virtual upward dispalcement at *B*.

δy_2 = Virtual upward displacement at *C*.

Virtual work done by the vertical reaction

R_B = Force × Displacement

$= R_B \times$ Virtual displacement

$= R_B \times \delta y_1$

∴ Virtual work done by the reaction,

$$R_A = R_A \times \text{Virtual displacement}$$
$$= R_A \times 0 = 0$$

Virtual work done by load

$$P = -P \times \text{Virtual displacement at } C$$
$$= -P \times \delta y_2$$

(Here minus sign indicates that the force P is acting downwards but virtual displacement is upwards)

$\therefore$ Total virtual work done on beam

$$= R_A \times \delta y_1 + 0 - P\delta y_2$$
$$= R_B \times \delta y_1 - P \times \delta y_2 \qquad (i)$$

The virtual displacement δy_2 is converted in terms of δy_1. This is done with the help of similar tringlés ABE and ACD as,

$$\frac{AC}{AB} = \frac{CD}{BE} \text{ or } \frac{a}{L} = \frac{\delta y_2}{\delta y_1}$$

$$\therefore \qquad \delta y_2 = \frac{a}{L} \times \delta y_1 \qquad (ii)$$

Substituting the value of δy_2 in equation (*i*), we get

Total virtual work done

$$= R_B \times \delta y_1 - P \times \frac{a}{L} \times \delta y_1$$
$$= \delta y_1 \left[R_B - P \times \frac{a}{L} \right] \qquad (iii)$$

But the algebraic sum of the virtual works should by zero from the principal of virual work. Hence equating equation (*iii*) to be zero, we have

$$\delta y_1 \left[R_B - P \times \frac{a}{L} \right] = 0 \text{ or } R_B - P \times \frac{a}{L} = 0$$

$$\therefore \qquad R_B = P \times \frac{a}{L} \qquad (iv)$$

Similarly, the vertical reaction at A can be calculated. The reaction R_A will be

$$R_A = \frac{P \times b}{L} \quad (v)$$

Alembert's principle and dynamic equilibrium

Alembert's principle applicable to plane motion

Let us assume a body of mass '*m*' is moving with a uniform acceleration '*a*' under the action of external force *F(*also known as effective force*)*. Then according to Newton's seconds law of motion, we can write

$$F = m \times a$$

The equation (*i*) can also be written as

$$F - ma = 0$$

From equation (*ii*), it is clearly understood that by applying a force - *ma* on the body will be in equilibrium as the sum of all forces acting on the body is zero. Such equilibrium is called *dynamic equilibrium* and the force 0- *ma* is called **D' Alembert force** or *reserved force of inertia force.*

The body will be in dynamic equililbrium under the action of external force *F* and the reversed effective force (or inertia force). This known as *D*' Alembert's Principle.

Hence *D' Alembert's principle states that the net external forces* acting on the system and the *resultant reversed effective forces (or inertia forces) equillibrium.* This is D' Alembert's principle applecabel to plane motion.

The inertia force on a body = mass of the body × acceleration..

Inertia force acts in the opposite direction of motion of the body and passes through the C.G. of the body. The above method is considerably quicker for problems where a number of bodies are connected at this method does not require the consideration of tension in the string.

To explain this principle, which is applicable to plane motion consider the two following cases :

1. *Analysis of the motion of two bodies connected by a string.*

Two bodies of weights W_1 and W_2 are connected to the two ends of a light enextensible string passing over a smooth pulley as shown in Figure 7.20. The weight W_1 is mort than weight W_2.

Let a = Acceleration of the system.

Now the net external force acting on the system in the direction of motion

$$= (W_1 - W_2). \quad (i)$$

Reversed effective force (or inertia force) on weight W_1

$$= -\text{ Mass of } W_1 \times \text{Acceleration}$$

$$= -\frac{W_1}{g} \times a \qquad \left(\because \text{Mass of } W_1 = \frac{W_1}{g}\right)$$

Reversed effective force on weight W_2

$$= -\text{ Mass of } W_2 \times \text{Acceleration}$$

$$= -\frac{W_2}{g} \times a$$

$\therefore$ Resultant reversed effective force

$$= -\frac{W_1}{g} \times a + \left(-\frac{W_2}{g} \times a\right) = -\frac{a}{g}[W_1 + W_2].$$

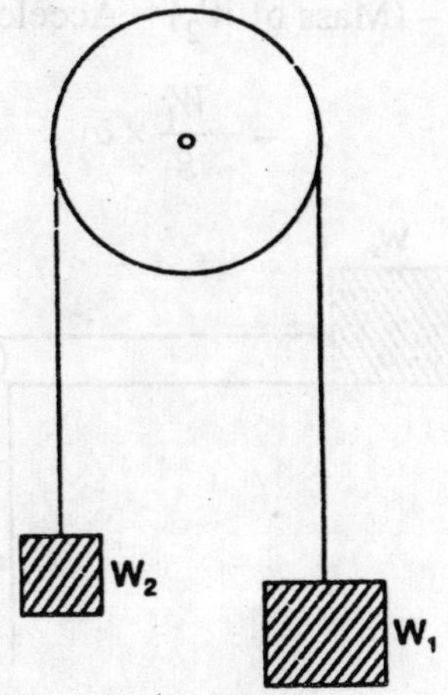

Figure 7.20

But the net external force acting on the system and the resultant reversed effective force should be in equililibrium in accordance with the D' Alembert's principle.

∴ Net external force + resultant reversed effective force = 0

or $$(W_1 - W_2) - \frac{a}{g}(W_1 + W_2) = 0$$

or $$(W_1 - W_2) = \frac{a}{g}(W_1 + W_2)$$

∴ $$a = \frac{g(W_1 - W_2)}{(W_1 + W_2)}$$

2. *Analysis of motion of two bodies connected by a string when one body is lying on a smooth horizontal surface and other is hanging freely.*

Let a = Acceleratrion of the system.

Net external force acting on the system shown in Figure 7.21 in the direction of motion = W_1

The weight W_2 is not acting in the direction of motion.

Reversed effective force (or inertia force) on weight W_1

$= -$ (Mass of W_2) × Acceleration

$$= -\frac{W_1}{g} \times a \qquad \left(\because \text{Mass} = \frac{W_1}{g}\right)$$

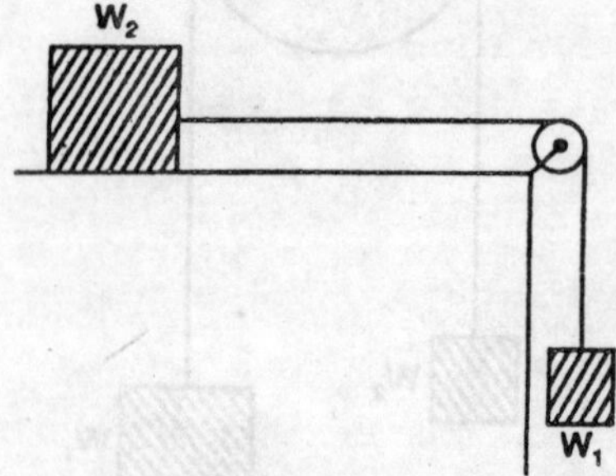

Figure 7.21

Reversed effective force on weight W_2

$$= -\text{(Mass of } W_2) \times \text{Acceleration}$$

$$= -\frac{W_2}{g} \times a.$$

(This force will be acting of W_2 in the opposite direction of motion of W_1 *i.e.* this force acts in the horizontal direction towards left of W_1 and will pass through the C.G. of the body of weight W_2).

∴ Resultant reversed effective force

$$= -\frac{W_1}{g}a + \left(-\frac{W_2}{g}a\right) = -\frac{a}{g}\left[W_1 + W_2\right]$$

But according to D'Alembert's principle, the net external force acting on th system in the direcion of motion and the resultant reversed effective forces should be in equilibrium

Net external force + resultant reversed effective force = 0.

or $$W_1 + \left[-\frac{a}{g}(W_1 + W_2)\right] = 0$$

or $$W_1 = \frac{a}{g}\left[W_1 + W_2\right] \quad \therefore \quad a = \frac{gW_1}{(W_1 + W_2)}$$

D' Alembert's principle applicable to rotary motion

It states that when external torques (also called active torques) acts on a system having rotating motion, then the algebraic sum of all the torques acting on the system due to external forces and reversed active forces including the inertial torqures (taken in the opposite direction of the angular acceleration) is zero.

To illustate the above principle, consider the two following cases :

1. *Where Rotation due to a weight W attached to one end of a string, passing over a pulley of weight* W_0

A weight W is attached to one end of an inextensible string, which passes over a pulley of weight W_0, the other end of the string is attached to the periphery of the pulley. The rotation to the pulley is caused in clockwise direction when the weight W moves downwards.

Let a = Linear acceleration of the weight W.

α = Angular acceleration of the pulley

R = Radius of the pulley

I = Moment of inertia of the pulley about the axis of rotation.

But we know

Linear accelearation = Anglular acceleration × radius.

$$\therefore \quad a = \alpha \times R$$

Reversed effective force of weight W

$$= -\text{(Mass of } W) \times \text{Acceleration}$$

$$= -\left(\frac{W}{g}\right) \times a = -\left(\frac{W}{g}\right) \times \alpha \times R$$

NowTorque due to external force = $W \times R$

Torque due to reversed effective force

= (Reversed effective force) × R

$$= \left(-\frac{W}{g} \times \alpha \times R\right) \times R = -\frac{W}{g} \times \alpha \times R^2.$$

Torque on the pulley due to angular acceleration is given by equation (7.20).

$$= I \times \alpha.$$

∴ Inertia torque, taken in opposite direction of the angular acceleration

$$= -I \times \alpha.$$

Then according to D' Alembert's Principle, the algebraic sum of all the torques on the system due to external forces and reversed active forces including inertia torques (taken in the opposite direction of the angular acceleration) should be zero.

$$\therefore \quad W \times R - \frac{W}{8} \times \alpha \times P^2 - I \times \alpha = 0$$

or $$W \times R = \frac{W}{g} \times \alpha \times R^2 + I \times \alpha = \alpha\left[\frac{W}{g}R^2 + I\right]$$

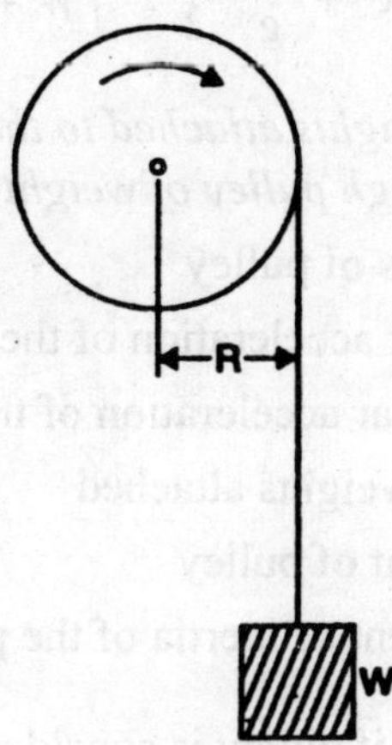

Figure 7.22

$$\therefore \qquad \alpha = \left(\frac{W \times R}{\frac{W}{g}R^2 + I}\right)$$

If linear acceleration 'a' is to be determined, then

$$\alpha = \frac{a}{R}.$$

$\therefore$ Equation (*i*) becomes as

$$\frac{a}{R} = \frac{W \times R}{\frac{W}{g}R^2 + I}$$

$$\therefore \qquad a = \frac{WR^2}{\frac{W}{g}R^2 + I}$$

If the pulley is considered to be a solid disc, then moment of inertia of the pyulley is given as

$$I = \frac{MR^2}{2}, \text{ where } M = \text{Mass of pulley} = \left(\frac{W_0}{g}\right) = \left(\frac{W_0}{g}\right)\frac{R^2}{2}$$

Substituting this value in equation (*ii*), we get

$$a = \frac{WR^2}{\frac{W}{g}R^2 + \frac{W_0}{g}\frac{R^2}{2}} = \frac{gW}{\left(W + \frac{W_0}{2}\right)}$$

2. *Rotation due to weights attached to the two ends of a string, which passes over a rough pulley of weight* W_0.

Let R = Radius of pulley

a = Linear acceleration of the system

α = Angular acceleration of the pulley

W_1, W_2 = Two weights attached

W_0 = Weight of pulley

I = Moment of inertia of the pulley

$= \frac{MR^2}{2}$ if pulley is considered a solid disc

$= \frac{W_0}{g}\frac{R^2}{2} \left(\because \text{Mass of pulley} = \frac{\text{Weight of pulley}}{g}\right)$

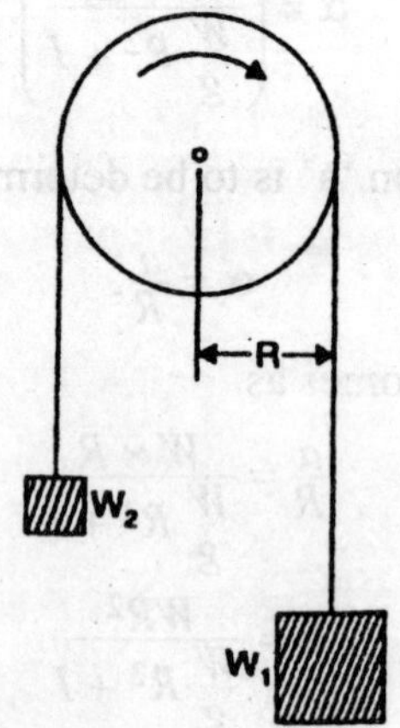

Figure 7.23

Considered $W_1 > W_2$, so thatthe rotation of the pulley is caused in clockwise direction.

Resultant torque caused by the weights

$$= W_1 \times R - W_2 R = (W_1 - W_2) \times R$$

Reversed effective force on weight W_1

$$= -\text{ (Mass of } W_1) \times \text{Acceleration } = \frac{W_1}{g} \times a$$

$\therefore$ Torque due to reversed force on weight W_1

$$= -(\text{Mass of } W_1) \times \text{Acceleration} = \frac{W_1}{g} \times a.$$

Reversed effective force on weight W_2

$$= -\frac{W_1}{g} \times a \times R$$

Torque due to reversed effective force on weight W_2

$$= -\frac{W_2}{g} \times a$$

Torque to angular acceleration on pulley = $I \times \alpha$

(See equation 7.20)

Inertia torque taken in opposite direction = $-I \times \alpha$

Now, the algebraic sum of all the torques acting on the system due to external forces and reversed effective forces inliuding inertia torques (taken in opposite direction) should be zero, in accordance with the D' Alembert's principle.

$$\therefore \quad (W_1 - W_2) \times R - \frac{W_1}{g} \times a \times R - \frac{W_2}{g} \times a \times R - I \times \alpha = 0$$

or $$(W_1 - W_2) \times R - \frac{W_1}{g} \times a \times R - \frac{W_2}{g} \times a \times R - I \times \frac{a}{R} = 0$$

$$\left(\because \alpha = \frac{a}{R}\right)$$

or $$(W_1 - W_2)R^2 = \frac{W_1}{g} \times a \times R^2 + \frac{W_2}{g} a \times R^2 + I \times a$$

$$- a\left(\frac{W_1 R^2}{g} + \frac{W_2 R^2}{g} + I\right)$$

$$a = \frac{(W_1 - W_2) \times R^2}{\left(\frac{W_1 R^2}{g} + \frac{W_2 R^2}{g} + I\right)}$$

$$= \frac{(W_1 - W_2) \times R^2}{\left(\frac{W_1 R^2}{g} + \frac{W_2 R^2}{g} + \frac{W_0}{g}\frac{R^2}{2}\right)} \quad \left(\because I = \frac{W_0}{g}\frac{R^2}{2}\right)$$

$$= \frac{g(W_1 - W_2)}{\left(W_1 + W_2 + \frac{W_0}{2}\right)}$$

Index